Acting in an Uncertain World

Inside Technology
edited by Wiebe E. Bijker, W. Bernard Carlson, and Trevor Pinch

Acting in an Uncertain World

An Essay on Technical Democracy

Michel Callon
Pierre Lascoumes
Yannick Barthe

translated by Graham Burchell

The MIT Press
Cambridge, Massachusetts
London, England

First MIT Press paperback edition, 2011

© 2001 Editions du Seuil.

This translation © 2009 Massachusetts Institute of Technology.

First published as *Agir dans un monde incertain: Essai sur la démocratie technique* by Editions du Seuil.

Set in Stone Serif and Stone Sans on 3B2 by Asco Typesetters, Hong Kong.

Publication of this book has been aided by a grant from the French Ministry of Culture.

Library of Congress Cataloging-in-Publication Data

Callon, Michel.
[Agir dans un monde incertain. English]
Acting in an uncertain world : an essay on technical democracy / Michel Callon, Pierre Lascoumes, Yannick Barthe.
    p. cm.
Originally published in French as: Agir dans un monde incertain.
Includes bibliographical references and index.
ISBN 978-0-262-03382-4 (hc : alk. paper)—978-0-262-51596-2 (pb. : alk. paper)
1. Democracy. 2. Political leadership. 3. Technology—Political aspects. 4. Science—Political aspects. I. Lascoumes, Pierre, 1948–. II. Barthe, Yannick. III. Title.
JC423.C245413 2009
321.8—dc22   2008021376

This book is dedicated to all those who, by inventing technical democracy, re-invent democracy.

Contents

Acknowledgements

(

Every book that originates in research is a collective work. This one is no exception to the rule.

This book owes much to several years of continuous discussions with our colleagues at the Center for the sociology of innovation at the École des mines in Paris. It has also been nourished by many exchanges between the authors and Marie-Angèle Hermitte, Pierre-Benoît Joly, Philippe Roqueplo, and Michel Setbon. But assuredly it would have been impossible without the work carried out by Janine Barbot and Nicolas Dodier, Phil Brown, Steven Epstein, Jacques Lolive, Sophie Houdart, Karin Knorr, Christian Licoppe, Vololona Rabeharisoa, Elizabeth Rémy, and Brian Wynne. We owe more than just references to these colleagues; the empirical material they have brought together and worked on forms the substance of this book. In a sense they are its co-authors. We especially thank Vololona Rabeharisoa and Cyril Lemieux, whose close and critical reading has made a considerable contribution to the clarification of our arguments.

The program on risk, led by Claude Gilbert with unusual steadfastness and effectiveness, has contributed powerfully to giving legitimacy to the debates around technical democracy; it has also provided an irreplaceable framework for discussion between specialists of the social and natural sciences, as well as between researchers and actors. In the preparation of this book we have also benefited from the help of Jacques Theys, Ministère de l'Équipement, Direction de la Recherche et des Affaires Scientifiques et Techniques.

Without the constant support of the École des mines de Paris, this work would never have seen the light.

Finally, how could we fail to mention all that we owe to the stimulating work of Bruno Latour? The present book is an echo of his *Politics of Nature* and the continuation of a dialogue that, for one of us, began 25 years ago, and which we hope will continue for many more years.

Acting in an Uncertain World

Prologue

Friday, 17 December 1999. French Prime Minister Lionel Jospin is simultaneously celebrating the end of the century, the New Year, and the renewal of ties between France and Japan after years overshadowed by the resumption of nuclear tests. In front of representatives of the French community assembled for the occasion at the French embassy in Tokyo, he embarks on the summation of his speech: "In my name, in the name of the French government and of the French people at home, I bring you my most sincere calves [veaux]." Surprise in the audience, and then mild amusement. The prime minister, who, like everyone else, knows from the great Sigmund that no action is more successful than those we call slips, immediately corrects himself: "No thoughts about mad cows will be admitted. Please accept all my good wishes [vœux]. There you are, this shows how weighty this issue is."

Not content with frightening European consumers and poisoning relations between France and England, the mad cow trips up a French prime minister on a foreign visit. This peaceable ruminant is suddenly transformed into a dangerous political animal that everyone should be wary of! Beware of the cows for they are no longer guarded!

By escaping from the enclosed pasture where it grazed in peace, the mad cow helped to spread the news that some had already had a premonition of for a long time: relations between science and power will never be the same. To make the right decisions, we thought, all we had to do was rely on indisputable knowledge. Now we must take decisions—no one can avoid doing so—just when we are plunged into the greatest uncertainty. What exactly are these prions that in a few months have become as famous as Saddam Hussein? What are they capable of doing? How far are they ready to go to make our life unbearable? An insidious, invisible enemy is amongst us. What is to be done when no indisputable fact or expert can reassure us? And as if there were only prions to torment us! The bustling whirl of radioactive waste, genetically modified organisms, and greenhouse gases give us sleepless nights.

The politicians are helpless. Some lose their heads, as if already affected by prions. In order to calm that new god, public opinion, an English Minister of Agriculture invites television cameras to witness the spectacle of his young daughter Cordelia biting into a British hamburger with gusto! How brave! But more to the point, had he taken care to get her to sign a statement of informed consent? In former times a king did not hesitate to sacrifice his daughter in order to placate the gods. But he had the decency, dare we say civility, to explain to her the gravity of the situation, indeed to convince her of the grandeur of an action that should save the country. Agamemnon is hard, but he hides nothing from Iphigenia, who ends up sacrificing herself for the common cause.

Every nation reacts in its own way. France with its slips of the tongue, England by playing Russian roulette for the media, and Japan—the Japan from which Lionel Jospin cannot hide his concerns—by importing procedures devised in the West for dealing with these difficult and increasingly numerous cases which mix together sciences, technologies, and societies without restraint, infinitely complicating the political decision makers' task.

The anecdotes that follow are drawn from Michel Callon's notebook.

We are no longer in Tokyo, but in Nara, a few kilometers from Kyoto. It is no longer the French Embassy, but a majestic conference hall in one of the most recent technopoles in Japan. As president of the Society for Social Studies of Science (4S), I have been invited to participate in a public symposium in which the conclusions of the first Japanese consensus conference on gene therapy are presented.

On the stage, several rostrums have been set up. Mr. Kiba steps up to the microphone and says:

The development of science and technology has a considerable impact on the lives of ordinary citizens. It gives rise to many new problems which are grouped under the heading of the social acceptability of technologies. These problems are raised in many domains, such as nuclear waste, the incineration of household waste, organ transplants, or even gene therapy. Political, economic, and ethical problems arise with regard to each of these issues. And it would be wrong to see these problems as secondary, or as separable from scientific and technical questions.

Kiba takes a breath, because he feels that the most difficult remains to be said:

Their formulation and resolution presupposes the direct involvement of citizens. But how can we ensure that laypersons, non-specialists, can give their views on technical subjects of such great complexity? Let us recognize, Kiba adds, that this cannot be left to the responsibility of existing political institutions. These were designed to protect the experts and not to allow the participation of non-experts.

Kiba breaks off. He seems alarmed by what he has dared to say. I have the impression that he is aware of the incongruity of his remarks. A Japanese giving public lessons on democracy? Now we have seen everything. I imagined the Japanese fixed on technical progress, concerned only with technological innovations. And here they are having uncertainties! However, if they ask questions that we imagined were reserved to Westerners, in the solutions they devise they are where we expect them to be: on the side of technology transfers, but in this case, the transfer of social technologies. The speaker continues:

In Europe, many experiments have been carried out in order to resolve the problem of the social acceptability of technologies through greater citizen involvement. We have made a careful inventory. One of the most interesting procedures seems to us to be the one devised by the Danes, which they call the consensus conference.

Kiba embarks on the history of this procedure. Invented in the United States, but applied there solely to the question of the definition of medical practice, it was taken up by the Danes, who transformed it profoundly. Kiba mentions that several countries have already been inspired by the Danish experience. He cites the United Kingdom, New Zealand, and the Netherlands. France is not on the list, because the citizens' conference on genetically modified organisms (GMO) will not take place until the following year in Paris.

A good Japanese who makes the cultural exception of Japan a constitutive feature of its culture, Kiba continues:

It is often said that Japanese culture does not lend itself to the organization of a democratic debate on technology. But this is not inevitable.

Kiba explains how the idea arose of organizing a consensus conference on gene therapy, an emerging and already hot subject that raises a number of ethical problems. He tells how the support of Toyota was obtained in order to make up for the lack of commitment from public authorities, and how it was decided to transform this first endeavor into an experiment. The aim, he emphasizes, was not to arrive at results that could be used, but to evaluate the procedure itself in order to figure out its limits and identify possible improvements. The Japanese are past masters of the art of transposition and enrichment, and they know that the adoption of technologies—including social technologies, as in this case—is above all a matter of adaptation.

Speakers follow one another to the rostrum, observing a regular protocol. One speaker gives a detailed account of how the panel of citizens was selected, how the training sessions and the question-and-answer exchanges with the experts were organized, then how the final proposals were drafted, and finally how this final session and the dialogue with a hand-picked but wider audience were constituted. This speaker ends his presentation with a commentary that demonstrates the extent to which the organizers have been able to distance themselves from the experiment they have conducted:

It is important to introduce ordinary citizens into the debate and to get them to participate in working out the measures that will be taken. But this is not an end in itself. The consensus conference is certainly a procedure that aims to increase the democratization of decision making, but this is not its only purpose. The content of the decisions it allows to be taken is not without importance. From this point of view, it should be compared with other, existing procedures.

It is precisely in order to facilitate the evaluation of this procedure that the organizers have asked some foreign figures to give their point of view both on the overall project of the democratization of decision making and on the procedure itself.

Now it is Sheila Jasanoff's turn to speak. Sheila was a professor at Cornell University, where she headed the interdisciplinary Science, Technology, and Society (STS) program, whose objective is to train students who will be able to take up the new cultural, political, economic, and organizational challenges posed by the increasing importance of the technosciences in our societies. Sheila, a jurist by training, is a recognized authority in our field. "The achievement of a half-hearted consensus," she states, "is the worst objective we could have in our complicated societies." She is insistent:

Agreement is often reached to the detriment of opponents or the recalcitrant who have been unable to express themselves or who have been silenced. And then agreement reached at a given moment may very well no longer be valid a bit later when the circumstances have changed. Agreement is only rarely desirable!

Sheila is right. Consensus is often a mask hiding relations of domination and exclusion. Democracy will not be increased by seeking agreement at any cost. Politics is the art of dealing with disagreements, conflicts, and oppositions; why not bring them out, encourage them, and multiply them, for that is how unforeseen paths are opened up and possibilities increased.

Now she comes to the procedure itself:

A consensus conference only has point when it is carried along by a wider current and is immersed in multiple, constant debates. Gene therapy has been discussed in the United States for twenty years, or rather, all the problems it raises either directly or indirectly, questions of intellectual property, of clinical experimentation, have been and continue to be debated in different institutions, commissions, forums, and by a multiplicity of groups and persons with very often divergent, indeed contradictory conceptions and interests.

Sheila seems to be telling the Japanese: "Democracy is not a gadget. It is not something you copy; it is not just a matter of a few procedures. It is something deeper that must seize hold of the social body at its very core."

As for the procedure itself, and independently of the conditions of its application, which, it is understood, do not convince the speaker, in her eyes it suffers from serious defects:

What is at stake in these procedures is that the professionals learn something from laypersons. Is this really the case here? I am not sure. And then, above all and first of all the procedure must result in some political decisions. Now permit me to be skeptical on this point, for your initiative was taken, as you have just said, outside of any governmental demand. It was supported by a private foundation. It is difficult to see it giving rise to any decision making. It is therefore a complete waste of time, a parody of democracy.

It is a harsh judgment. But why should the social sciences be soft? When Sheila finishes her talk, silence fills the hall and its monumental architecture suddenly seems glacial. However, the symposium's procedure quickly moves things along. It is the turn of the panelists, and then the experts, to give their views. The latter are still suffering from the shock of their experience. One of them summarizes the general opinion: "I was skeptical. I now think it is necessary to accompany research and to organize this kind of discussion."

The ordinary citizens are no less satisfied. They avow that their position with regard to gene therapy is much more reserved than it was before the conference. But debate becomes possible, as one of them summarizes magnificently: "Thanks to the conference I have become an amateur of gene therapy. And as an amateur, there are things that I like, and others that I am less keen on."

We are familiar with the strange movements between the West and the Far East, and the game of well-oiled roles to which they give rise. The West shows the way, like the Statue of Liberty holding out the flame of liberty to the rest of the world, and Japan, needy and assiduous, is supposed to follow. The Japanese are past masters in the art of playing this role, which allows them both to preserve their identity (they are different) and to readily share in a common history (they copy). The role playing requires that the Japanese, having imitated the model, hasten to surpass it and give lessons to their old teachers.

San Diego. The annual colloquium of 4S. More than 500 researchers from all over the world. The Japanese are there. Some have suggested organizing a session on consensus conferences. The theme has never previously been taken up at our gatherings. No doubt it was considered to be too applied, too close to the daily concerns of decision makers! Our Japanese colleagues are not paralyzed by these misgivings. They give a detailed presentation of the two Japanese experiments. (After the conference on gene therapy, another conference was organized on information technologies and on the Internet in particular.) They reveal what we had only briefly glimpsed at Nara: Five researchers from STS were behind the first conference. Reading the literature, they had come across the Danish experiments.

Kobayashi, one of the speakers, gives a detailed description of the two conferences. He demonstrates his absolute familiarity with experiments conducted throughout the world. A good professional, he explains critical points of the procedure, including the recruitment of members of the panel, the choice of experts, the duration of training, the format of the final proposals, and the right of expression for minority points of view. Then he comes to the lessons he thinks can be drawn from this experience:

It has often been claimed, and what's more continues to be claimed, that scientific and technical questions are too complicated for laypersons to be able to make sensible judgments. And, once again, the miracle, which is no longer a miracle moreover, took place: all the specialists were surprised by the quality of the final documents.

Kobayashi wonders:

What is it in the production of laypersons that surprises the specialists?

For him, what is surprising is that the laypersons, these amateurs of gene therapy, were perfectly capable of assimilating the technical details, but they also helped to enrich the experts' knowledge:

One episode was particularly illuminating. A clinician participating in the conference as an expert provided the panel with copies of a document given to patients in order to get their informed consent. This document, he explained, was carefully worked out and tested and he was confident of its quality. However, much to the surprise of the clinician, the panel found it of very mediocre quality. The ordinary citizens stressed the degree to which the document, peppered with technical terms, each more obscure than the other, was incomprehensible to a patient who had to decide whether or not to take part in an experiment. What is more, one of the panel members pointed out to the clinician that the phrase concluding a section of the document was, to say the least, shocking. In fact one could read: "If the therapy has an unfortunate outcome, we would be very grateful if you were to bequeath your body to medicine."

One of the qualities of a specialist is to think of everything! Kobayashi continues:

This anecdote illustrates the complementary relationship between knowledge produced in the laboratory and its conditions of utilization.

Fearing that we had not grasped the significance of his remarks, Kobayashi recounts the particularly illuminating comment of a Japanese chemist:

This great scientist said that from now on chemistry must be able to complete the list of the properties of molecules in the laboratory and to enrich this list with the characteristics of these same molecules, but taken outside the laboratory.

Spot on! Laboratory research and research outside the laboratory: we should have thought of this obvious symmetry ourselves. Molecules do not live only in the closed space of the laboratory or in places that reproduce the conditions of the laboratory. They also move around in the open! That is where ordinary citizens are waiting for them, observe them, and strive to control them. Hence consensus conferences, public hearings and inquiries, and focus groups.

The session is drawing to an end. Kobayashi continues, imperturbable:

Can we introduce procedures for not only consulting citizens but also for involving them in the production of knowledge on issues that provoke confrontations which, as in the case of nuclear power plants, have become more serious in recent years? How can we ensure that the proposals and conclusions produced by citizens' panels are taken into account in public decisions?

Kobayashi comes to the end of his presentation. He cleverly returns to its title: Who has most to learn, experts or laypersons? The answer follows logically from his remarks: "Obviously, the experts!"

On the flight back to Paris, I come across an article in a magazine written by a colleague. He draws some lessons from the citizen conference in June 1998 organized by the Parliamentary Office for the evaluation of scientific and technological options. He says rightly that after this experience nothing will be the same. A landmark has been passed—one as symbolic as Cape Bojador, on which Portuguese sailors came to grief long ago, the way to the Indies being open to them once they had passed it. For some weeks the public space has been invaded. Genetically modified organisms have left the research centers where they were confined. They have had a good time marching with angry farmers, spreading through magazines, speaking to the evening television news programs through the ordinary citizen, and arousing controversy. As predicted for a long time, they were finally there in our midst. They were there, but not in hiding, and not invisible and discreet as some would have liked. No! They were showing themselves without false modesty, proudly riding high in the media. Whatever its obvious limits, this colleague added, the citizen conference, for a time at least, had made visible and debatable what had been hidden and excluded from public debate.

It is true that there was something euphoric about the chaos that was organized in this way. José Bové, a very popular leader of a leftist farmers' trade union, revived the social movement, dragging in his wake intellectuals, sociologist-journalists, and journalist-sociologists who no longer believed in it. Experts multiplied in front of the cameras to say that they were not as positive as some would like it to be thought and that these debates had their good points. One sententiously discoursed endlessly on the principle of precaution; all of them put in their warnings and interpretations. "Let's decide!" said some. "Yes, that's it, let's settle it!" said others. "Above all let's not lose time!" added anxious economists. "Can't you see that the Americans are profiting from it to conquer the market?" "Let's take our time," murmured the calmest. "Let's not be beguiled by powerful interest groups; let's consult and deliberate."

The citizen conference helped bring it about that technological progress was once again debatable, and that the market ceased being that obscure force, or deliberately obscured force, which dispenses with all political deliberation. Even the French Academy of Sciences, in its "great wisdom," heard the message. Without delay it got in line with current tastes, organizing forums on the health consequences of mobile phones, or on the effects of dioxins, though not long ago it had been happy to say "Move along, there's nothing to see, all these rumors are the fruit of a sick collective imagination, of an unconscious fear that seizes hold of the people when new technologies appear." And not long ago the French Academy of Sciences would have been happy to recall the long list of irrational resistances that have marked the history of industrialized societies: Remember the

Luddites, the machine-wreckers! Remember the railway and the ridiculous fears it aroused! Remember! Remember!   ∎

Let us remember above all Kobayashi and his modest conclusions. Science and technology cannot be managed by the political institutions currently available to us. Obviously, it is not a question of dismantling them. They have given ample proof of their effectiveness. But their limitations are no less obvious. They must be enriched, expanded, extended, and improved so as to bring about what some call technical democracy, or more precisely in order to make our democracies more able to absorb the debates and controversies aroused by science and technology.

GMOs, BSE, nuclear waste, mobile phones, the treatment of household waste, asbestos, tobacco, gene therapy, genetic diagnosis—each day the list grows longer. It is no good treating each issue separately, as if it is always a case of exceptional events. The opposite is true. These debates are becoming the rule. Everywhere science and technology overflow the bounds of existing frameworks. The wave breaks. Unforeseen effects multiply. They cannot be prevented by markets, any more than by the scientific and political institutions. It was thought that genetic diagnosis kits had been perfected without a problem, and now some cry blue murder; the pursuit of profit, they maintain, leads straight to eugenics. We thought that geology would ensure a decent and definitive burial for nuclear waste that everyone would respect, and now wine growers, whose voice had not been heard, are worried, not about the effects of radioactivity, but about far more worrying commercial effects, since they are in danger of losing foreign customers who could take fright on learning that the grapes ripen some hundreds of meters above containers filled with nuclear substances!

It would be pointless to erect barriers to contain these overflows; they would quickly give way one after the other. First of all we should recognize that these overflows are destructive only if we stubbornly seek to prevent them. When given the space they need, they reveal their fecundity, their fertilizing power. In chapter 1 we endeavor to demonstrate what this power to enrich political debate consists in by emphasizing the importance of collective experimentation and learning. In *hybrid forums*, in which the direction given to research and the modes of application of its results are discussed, uncertainties predominate, and everyone contributes information and knowledge that enrich the discussion.

These overflows make it clear that the great divisions are outmoded. As Kobayashi rightly said, to start with we should accept the fact that the knowledge of specialists is not the only knowledge possible, and consequently we should recognize the richness and relevance of knowledge developed by laypersons, and in particular by the groups that these overflows directly or indirectly concern. The conviction (both in minds and in institutions) that there is a difference in kind between the knowledge developed by professionals and that developed by laypersons is so strongly rooted that we will need at least two chapters to establish a new parity! Chapter 2 shows what *secluded research* consists in, that is, laboratory research which is not ruled out, but overflowed, when the molecules and genes it studies are let out in the open. Secluded research risks paralysis if it refuses to cooperate with *research in the wild*. In chapter 3 we present the characteristics of research in the wild and the modes in which it collaborates with laboratory research with the aim of getting the measure of overflows.

The *raison d'être* of the many procedures that have been invented and tried out over the last 30 years in all the so-called developed countries is that of organizing and controlling overflows, but without seeking to contain, prevent, or eliminate them. The consensus conference is only one of the apparatuses that have been devised to come to the aid of existing institutions. There is now a whole battery of procedures available for organizing hybrid forums. Chapter 4 shows that, in their diversity, they can be analyzed according to two dimensions. The first is the intensity of cooperation they establish between secluded research and research in the wild. The second is the amount of space they leave open for the emergence and consideration of new groups and new identities, whether it is those living near a nuclear power plant, parents affected by the death of their children, or patients who seek to participate in drug trials.

Chapter 5 presents some of the different existing procedures, showing how each enriches the scientific and political institutions in its own way. A democracy comes into play that can be described as *dialogic*. By absorbing the uncertainties that it puts at the center of debate, dialogic democracy enriches traditional representative democracy, which we propose to call *delegative* democracy.

Chapter 6 pursues the work of investigation of experiments underway by showing the consequences they entail for the notion of political decision making. In the space of organized hybrid forums, collective learning, which simultaneously produces new knowledge and new social configurations, ends up fabricating a close weave of micro-decisions, each of which is subject to discussion and linked to those that precede it as well as those that follow. This favors options being kept open instead of being quickly, and often irrevocably, closed down. The model of the clear-cut decision disappears along with the oft-repeated myth of Alexander drawing his two-edged sword to cut the Gordian knot that no expert managed to untie. Sheathe your swords! This is the slogan that could sum up the now-famous principle of precaution. No more clear-cut, bloody decisions. Manly warrior assurance is not replaced by inaction, but by *measured action*, the only possible action in situations of high uncertainty.

Measured action gives notice to a whole series of notions and oppositions of which the reader will find no trace in this book: nothing on risks, nothing recalling the distinction between fact and value, or between nature and culture, and nothing that reinforces the idea of omnipotent laws of the market. In chapter 7 we show that the effect of all these notions is to divert our attention and dissuade us from taking seriously all the endeavors to go further than the habitual procedures of consultation and representation. This suggests to us, in conclusion, that, by inventing the concrete modalities of a democracy that can pick up the challenge of the sciences and technologies, all the anonymous actors who have modestly devoted themselves to opening up new sites and experimenting with new procedures have contributed to the more general, never-completed enterprise of the *democratization of democracy*—that is to say, of the people's control of their destiny. There is a paradox in this: the philosophy in the wild practiced by the Danes or the Dutch is every bit as valid as all the confined moral and political philosophies that we find surfeit of on campuses and in other closed spaces.

1 Hybrid Forums

In March 1987, at intervals of a few days, the same scene takes place in the rooms of the prefectures of four French departments. Dozens of local councilors, mayors, and departmental councilors attend a "briefing." The prefect who has called them together has not clarified the purpose of the meeting, but their presence seems to be of the greatest importance. Proof of this is the diligence shown by the prefecture services. The summons was sent the previous day by telegram, and police cars have been sent to facilitate the councilors' movement.

During the meeting, the prefect quickly hands over to officials from ANDRA. ANDRA? The participants, who have never heard this strange acronym, learn that it is a national agency created within the Commissariat à l'énergie atomique (Atomic Energy Commission) with responsibility for radioactive waste.[1] It is this task that explains their presence in the various departments. "To eliminate certain nuclear waste that will have significant radioactivity for several thousands of years, burying it in deep geological strata has been considered," one of the experts from Paris explains. In a slightly professorial tone, he adds: "Inasmuch as some of these geological formations have been stable for millions of years, we assume that they will continue to be so for the period of decrease in radioactive elements. The geological structure will constitute then a 'trap' more than 400 meters deep. This trap should enable the waste to be isolated from the environment when the containers have been destroyed by erosion and the memory of the site has been lost. This 'geological safe' offers an immense advantage: it makes all the always uncertain conjectures on the evolution of society pointless." The audience can only be reassured. Never mind the schemes of future generations that everyone has been talking about for some months. It matters little whether or not they take care of this difficult inheritance. What matters now is not the behavior of changeable human beings, but the long-term behavior of geological formations that are *a priori* favorable. Precise and technical questions take the place of vague and general preoccupations. In order to answer these questions it is enough to ascertain the quality of the accommodating rock and to develop the soundest possible predictive models. "A series of geological explorations will be undertaken on four sites chosen for their subsoil. At the end of these explorations, a single site, one meeting all the requirements, will be selected for the installation of an underground laboratory. It goes without saying," the scientists conclude, "that a project like this would be a source of jobs and of not inconsiderable earnings for the department in which it is situated."

The news spreads in a few hours. It has the effect of a thunderbolt in the four departments concerned. Residents, whom it had no doubt been forgotten to invite to the briefing, quickly form associations. They are opposed to what they see as a fait accompli, and they demand information on the project. Is it reasonable to bury nuclear waste irreversibly? Can we trust the studies of the geological explorations? Are there other solutions? In the villages of the Ain, the Maine-et-Loire, the Deux-Sèvres, and the Aisne, the four departments affected by these geological drillings, ANDRA organizes dozens of briefings and distributes hundreds of leaflets presenting the project. Communication specialists explain, popularize, and reassure. Thinking that these populations are in the grip of irrepressible fears and terrors, they proclaim *urbi et orbi* that there really is no risk. Or, they admit reluctantly, it can involve only a very small risk, in the distant future, at a time beyond our imagination. In any case, they add, there is no other solution. We really have to get rid of nuclear waste once and for all! We cannot pass on this heavy burden to our descendants! Burial is a technical necessity. It is also a moral duty with regard to future generations.

But ordinary citizens have learned to mistrust information provided by nuclear agencies, even when they seem to be above suspicion technically and morally. Ordinary citizens still remember the Chernobyl cloud, which the established experts dared to maintain would halt at France's borders. This is why they prefer to turn to other sources of information. Some figures of nuclear counter-expertise are invited to give their point of view on the ANDRA project. Discussion points gradually emerge. These specialists qualify the idea that geological storage is the only conceivable technical solution. In the heat of the controversy, the residents realize that there are many uncertainties and that the burial of radioactive waste is only one line of research, requiring lengthy and complex scientific studies. They also discover that, in the past, other solutions had been considered which, for reasons that are far from clear, were quickly abandoned without thorough investigation. There is the technique of transmutation, for example, which, by ensuring the destruction of radionuclides with a long life, would have the advantage of considerably reducing the uncertainties inherent in geological storage.

Awareness of the existence of these scientific and technical uncertainties leads to the reformulation of the terms of the problem and the emergence of new questions and new scenarios. What if future generations were to find more satisfactory methods for dealing with these burdensome residues? What if the technical capabilities of our distant descendants were to make it possible one day to develop this waste? And what if the irreversibility of storage was contrary to the scientific approach? And...?

Questions that were thought to have been settled definitively are reopened. Arguments multiply and the project constantly overflows the smooth framework outlined by its promoters. In the course of the controversy, unexpected connections are established between what should have been a simple technical project and a plurality of stakes that are anything but technical. Thus we see new actors taking up the problem, imposing unexpected themes for discussion, and redefining the possible consequences of the project. The Bresse poultry farmers, for example, point out a danger that the technicians, obsessed with the seismic and hydro-geological data concerning the department's subsoil, clearly could not imagine. This is the threat posed to the economic health of the regions concerned by the introduction of a center for storing nuclear waste. The relationship established in the consumer's mind between the quality of certain agricultural products and the presence of radioactive waste makes the farmers fear that the image of these products will be damaged. Seen by its promoters as a source of local economic development, the storage of nuclear waste becomes a potential threat to some commercial interests. Local councilors leap to the defense, anxious to defend the interests of their electors and restive at the imposition of a definition of the general interest that disregards local realities. They call for a national debate, for a pluralistic expertise, and for a better consideration of the social and economic aspects of the problem.

The conflict grows acrimonious and turns into a pitched battle. No one talks now of the risks associated with storage strictly speaking, but of the risk of riots on the part of what are deemed to be uncontrollable minorities. Soon, squads of the riot police are sent to protect the ANDRA technicians so that they can continue their work. At the same time, demonstrations increase, attracting more and more people. The inhabitants of the departments are intent on resisting, with violence if necessary, the arrogance of the technicians and the arbitrary decisions of the central power that deny the identity of their territory. To put an end to this climate of civil war, in

1990 the government decides to backpedal and declares a moratorium on the research being conducted by the ANDRA. The time has come for a complete re-examination of the case. Space is made for consultation with all the interested parties. Caught unawares, the government discovers the existence of institutions that could be useful to it. It seeks help from the College for the Prevention of Technological Risks and from the Parliamentary Office for the evaluation of scientific and technological choices. The first real French law concerning the nuclear domain, the law of 30 December 1991, called the "Bataille law" after the name of its rapporteur, arises from these consultations and discussions. This text, and the apparatuses it sets up, strives to open up the "black box" of science in order to promote a program of research justified by an uncertainty that is now acknowledged and accepted. The dominant feature is the refusal of a definitive choice, which is put back and will require a new law to be passed. In the meantime, it is envisaged that three major lines of research will be explored and regularly evaluated by a commission of independent experts and the Parliamentary Office for the evaluation of scientific and technological options. The political dimension of the issue is recognized. It is no longer a matter of identifying and negotiating risks, as in a contract between insurer and insured, but of establishing constraining procedures for managing the apparent contradiction between minority points of view and what some consider to be the general interest. Furthermore, the law introduces a new conception of the mode of political decision making. It is no longer a matter of deciding on the basis of indisputable scientific facts. The law outlines the framework of a gradual approach that favors adjustments and corrections. In a word, it is decided not to decide, but to take time to explore conceivable options before deciding.[2]

Let us change the scene, or the department rather. Let us leave the Bresse region and move to Sarthe, following in the footsteps of the sociologist Élisabeth Rémy.[3] The problem here is not the burial of nuclear waste but a high-voltage line installed by Électricité de France, or more precisely the effects of the electromagnetic fields produced by this line. For some time, in fact, strange phenomena have been occurring in a small rural commune, to the extent that its inhabitants feel like they are involuntary actors in a science fiction film. Sometimes it is the siren of the commune's fire truck that goes off on its own. At other times, despite many visits from the people who installed it, an automatic gate pleases itself and opens without being given the order. The inhabitants complain of frequent headaches and insomnia. Those who prided themselves on their iron constitution are frequently ill. There is said to be a child who is constantly pulling his hair out

... except when he goes on holiday—that is, when he moves away from the accursed village. It is also said that the suicide, leukemia, and cancer rates are increasing in the area, following, as if by chance, the track of the high-voltage lines. Faced with what they see as threats, the inhabitants organize, try to make a list of all these cases, and aggregate the multiple isolated facts produced over the whole of the territory in order to give consistency to the hypothesis of the harmful effects of electromagnetic fields on health. Others appeal to experts whom they judge to be independent in order to make measurements in their property and prove the danger. Their suspicion is encouraged by the ambiguous discourse of Électricité de France officials, who, while refusing to state publicly that there is no danger, consider that if there is a risk it can only be slight and, in any case, the problem is being studied.

Actually, the problem is being studied. The question of harmful effects of low frequency electromagnetic fields is keenly debated by specialists. Despite much epidemiological and biological research on the subject over 20 years, there are still many uncertainties. The hypothesis of a danger linked to exposure to low frequency electromagnetic fields from electric lines was raised seriously for the first time in 1979. That year, in the very official *American Journal of Epidemiology*, an American researcher published the results of a study showing a statistical relationship between cancers in children and exposure to electromagnetic fields. Since then investigations have been carried out aiming either to support or refute this hypothesis. But no certainty succeeds in settling the debate, and the experts are practiced in evasive answers. We cannot completely exclude the existence of a danger, they say; on the other hand, nothing permits proof of the contrary.

It has to be acknowledged that the problem posed is not an easy one to solve. Research aiming to identify possible danger comes up against difficulties that are confronted by every epidemiological study of effects produced by weak exposure to a substance deemed to be harmful. In these tricky cases several conditions have to be met before a sound diagnosis can be given. First, we must be able to identify precisely the populations affected and, consequently, we must be able to define a level of exposure above which given individuals are considered to have been exposed. Second, given that what is being researched are long-term effects, in order to get reliable results there should be an epidemiological follow-up of the population over several years. The third condition concerns the characterization of effects produced by low doses. Since it is difficult to apprehend these effects directly, hypotheses have to be formulated and widely discussed. A fourth uncertainty concerns the way in which what is called a dose of electromagnetic field is calculated: Should we accept the average accrued intensity of the exposure, the peak of exposure, its temporal variation, or its frequency? As can be seen, the experts and the groups concerned are faced with what may be described as radical scientific uncertainties. They are especially uncertain since there are some who have an interest that they are and ... that they remain uncertain. Imagine the predicament of Électricité de France if the danger were to be proven!

There are striking similarities between the two cases just set out. In the example of radioactive waste as in that of high-voltage lines, the uncertainties concerning the dangers incurred (whether long-term or short-term) are patent. In both cases, despite these uncertainties, indeed because of them, decisions nevertheless have to be made, or, as we say, "something must be done." In the two cases, the controversies bear at the same time on the characterization of the dangers and on the procedure to be established so as to arrive at what may be considered a credible and legitimate characterization. In both cases, the controversies take place in public spaces that we propose to call *hybrid forums*[4]—forums because they are open spaces where groups can come together to discuss technical options involving the collective, hybrid because the groups involved and the spokespersons claiming to represent them are heterogeneous, including experts, politicians, technicians, and laypersons who consider themselves involved. They are also hybrid because the questions and problems taken up are addressed at different levels in a variety of domains, from ethics to economic and including physiology, nuclear physics, and electromagnetism.

This kind of socio-technical controversy is on the increase. In this book we will visit some of the many hybrid forums that the unpredictable and often chaotic development of science and technology has created: the Mad Cow forum, that of genetically modified organisms or of avian influenza, the AIDS forum, and that of neuromuscular diseases or nanotechnologies. But before going further into the analysis of these controversies and their organization, dynamic, and possible closure, we propose to show that they are an appropriate response to the increasing uncertainties engendered by the technosciences—a response based on collective experimentation and learning.

Uncertain Times

Contrary to what we might have thought some decades ago, scientific and technological development has not brought greater certainty. On the contrary, in a way that might seem paradoxical, it has engendered more and more uncertainty and the feeling that our ignorance is more important than what we know. The resulting public controversies increase the visibility of these uncertainties. They underscore the extent of these uncertainties and their apparently irreducible character, thereby giving credit to the idea that they are difficult or even impossible to master. These uncertainties are most striking in the domains of the environment and health, undoubtedly the most fertile terrains for socio-technical controversies. In view of their role in the constitution of hybrid forums and their capacity to render the future opaque and threatening, is it not advisable to ask "What exactly are we talking about when we evoke the notion 'uncertain'?"

From Risk to Uncertainty
Let us be careful not to confuse the notion of uncertainty with that of risk, which is its false friend. The two notions tend to be used interchangeably in current language, but they cover very different realities.

The term 'risk' designates a well-identified danger associated with a perfectly describable event or series of events. We do not know if this event or series of events will in fact take place, but we know that it *may* take place. In some cases, statistical instruments applied to series of systematic observations performed in the past make it possible to calculate the event's probable occurrence, which will then be described as objective probability. In the absence of such observations, the probabilities assigned depend on the points of view, feelings, or convictions of the actors; these are called *subjective* probabilities. Whether objective or subjective, these probabilities have in common their application to known, identified events that can be precisely described and whose conditions of production can be explained.

The notion of risk is closely associated with that of rational decision. In fact, in order for such a decision to be made, three conditions must be met. First, we must be able to establish an exhaustive list of the options open to us. In the case of the management of nuclear waste, this implies that we can guarantee that the three strategies of deep burial, transmutation, and surface storage are the only strategies worth considering. Second, for each of the options under consideration, the decision maker must be able to describe the entities constituting the world presupposed by that option. In the case of deep burial, for example, we will consider a world made up of clay strata or granitic massifs, of groundwater, of heedless human beings, and of a terrestrial atmosphere that is inexorably warming. Finally, the assessment of the significant interactions that are likely to take place between these different entities must be feasible. Human beings may decide to sink mines, penetrating the geological safe unawares; equally, predicting a tidal wave linked to global warming, they may decide to bury their dwellings, which will then be exposed to water containing radioactive substances. If these three conditions are satisfied, then the decision maker can make comparisons between the options on offer. To account for this truly exceptional situation, decision theorists introduce a notion that will be very useful for us: that of possible states of the world. A state of the world is defined first by the list of human and non-human entities that make it up, and then by the interactions between these entities. In choosing a state of the world, we choose not only the entities with which we decide to live but also the type of history we are prepared to share with them. We refer to *possible* states of the world because we know of causal chains that could produce them. Another way of talking about these states of the world is to employ the notion of scenario, a notion dear to futurologists.

The notion of risk is indispensable for understanding the choices made by a decision maker. For a moment, let us entertain the evidently implausible hypothesis that the management of nuclear waste can be reduced to this analytical framework. If we follow this procedure, we will be led to distinguish a state of the world (or a scenario) in which the waste is buried deep, another in which it is transmuted, and a third in which it is stored on the surface. On the basis of the knowledge available to us, we will try to describe the significant interactions that may occur in each of these scenarios, especially those between the social world and the waste. In this way we will identify potentially dangerous events for certain social groups. Being able to predict developments and identify effects, the decision maker will thus be in a position to make a rational choice. Obviously this will depend upon his preferences and those of the actors he thinks must be taken into account. It will also depend, and this is the important point, on how the decision maker assesses the possible dangers associated with each scenario, and, in particular, on his calculation of the probability of their occurrence. The notion of risk plays a crucial role, therefore, in rational decision theory and in the choice between several possible states of the world that it presupposes. That is why, to avoid ambiguities, it is sensible to reserve use of the notion to these completely codified situations.

Let us agree to speak of risk only in those quite specific cases where the exploration of possible worlds (or, if you prefer, the establishment of conceivable scenarios) has been completed, revealing the possibility of harmful events for certain groups. We are completely familiar with these events and know the conditions necessary for them to take place, even if we do not know whether they will in fact occur, and even if all we know is the probability of their occurrence.

It is easy to see why the notion of risk, thus defined, does not enable us to describe situations of uncertainty or to account for the modes of decision making in such contexts. In actual fact, science often proves to be incapable of establishing the list of possible worlds and of describing each of them exactly. This amounts to saying that we cannot anticipate the consequences of the decisions that are likely to be made; we do not have a sufficiently precise knowledge of the conceivable options, the description of the constitution of the possible worlds comes up against resistant cores of ignorance, and the behavior and interactions of the entities making them up remain enigmatic. The conditions required for it to be relevant to talk of risk are not met. We know that we do not know, but that is almost all that we know: there is no better definition of uncertainty. In such situations the only option is questioning and debate, notably on the investigations to be launched. What do we know? What do we want to know? Hybrid forums help to bring some elements of an answer to these pressing questions.

Uncertainty is a useful concept because it prevents us from confusing hybrid forums with situations of risk. It is nevertheless a fuzzy concept covering diverse configurations. Obviously, uncertainties may be more or less radical. There is a vast space between dismal ignorance and an impeccable knowledge of the states of possible worlds. It is worthwhile plotting its contours, for that is where the hybrid forums install themselves. One way of realizing this cartographic work is to review the different forms of uncertainty and note the particular controversies to which each of them may give rise.

Radical Uncertainties

The most revealing examples of the situation of radical uncertainty correspond to what are called *development risks*. These are situations linked to the commercialization of substances whose dangers must be unknown to the producer when he puts them on the market. This case is all the more striking as these problems often concern products, like drugs, requiring authorization to be put on the market, which presupposes prior and public checking of their harmlessness. If harmful effects become apparent, it is only after several years, and their explanation will necessitate further delays. The most famous example is distilben, a drug that was widely prescribed in the 1950s for woman likely to miscarry. Not until much later was it realized that, if the product had no direct harmful effect on the mothers, it nonetheless triggered serious disorders in the children. These effects only became apparent at puberty (malformations of the reproductive apparatus, sterility, cancer). There was, therefore, a gap of 15–20 years between absorption of the product by mothers and the first clinical signs for their daughters. It took a long time to identify the latter. And it took even longer to establish that they had a common source in the treatment prescribed to the mother. The set of processes was reconstructed only at the end of the 1970s.

Another recent example is that of infected blood. Until 1983, when the first hypotheses of exposure to danger were formulated, hemophiliacs and people having blood transfusions were given dangerous, indeed mortal health-care products, the dangerousness of which, and how serious the danger was, no one had been able to predict.

In these kinds of situation, uncertainties can only be lessened *a posteriori*. That is why they deserve to be called radical. The question that arises in these conditions is clearly whether the dangerous nature of the substance could and should have been seen earlier. The answer is undoubtedly positive. Being able to anticipate and track down potential overflows, establishing a system of supervision, and systematically collecting data in order to sound the alarm as soon as bizarre events occur entail a long list of measures. This suggests that ignorance is not inevitable, and that to think in terms of uncertainty is already to provide oneself with the means to take its measure. Moreover, the courts share this conviction when they try to find those responsible. Justifications that "it is just bad luck" are less and less admissible. Hence the importance of emergent controversies, even and especially if they are aroused by prophets of calamity. History has taught us that Cassandra was not always wrong.

The Era of Suspicion

Opacity dissolves gradually, and situations of uncertainty in which the hypothesis of a danger emerges are distinguished from each other by the precision of observations and explanations.

We will talk of "plausible potential danger" when persons or life environments suffer damage that is perfectly describable but whose causes and precise nature remain unknown. Such situations often lead to the drawing up of inventories. Some actors embark, individually or collectively, on the collection of cases that may confirm the existence of a new threat. The uncertainties surrounding them encourage the informal and sometimes wild development of hypotheses that are not yet verified and are often not immediately verifiable. Controversy focuses on plausible but fictional scenarios that provide acceptable interpretations of the observed facts. Those who sound the alarm, whether laypersons or experts, are at the center of the debates.

The publication in the *British Medical Journal* of a study by the French epidemiologist J.-F. Viel on cases of leukemia in young children living near the French nuclear reprocessing plant at La Hague sparked a controversy that illustrates perfectly this entry into the era of suspicion. According to Viel, there are convincing arguments that allow the supposition that the observed connection between certain customs of the inhabitants (swimming, eating shellfish) and an atypical level of cases of leukemia (four observed cases rather than the expected 1.4) could be due to the presence of radioactive substances in the environment. It will take two successive expert commissions to pacify the public controversy and provide data acceptable to all the parties involved.

Suspicions do not ineluctably lead to studies concluding that there is no danger. In the case of the possible carcinogenic effects of mobile telephones, we see an impressive spread of works based on very different methodologies. In May 2000, one of the most respected scientific journals, *Nature*, published an article by De Pomerai et al. demonstrating the effects on worms of prolonged exposure to radiation weaker than that emitted by mobile phones. Biological changes (the appearance of specific proteins) are observed that are analogous to those usually triggered by thermal stress. In view of the constant character of this type of response to heat, the authors consider that comparable phenomena are conceivable in the human being. These results conflict with others, which are more reassuring, but based on studies financed, at least partly, by the manufacturers. As a "precaution," the British government recommends a maximum restriction of the use of mobile phones by children, in view of the consideration that their developing nervous system is likely to make them highly vulnerable. These preliminary works led to the launch in the summer of 2000 of a major epidemiological campaign by the International Agency for Research on Cancer (IARC). Its aim is to identify several thousands of cancer cases (brain tumors, cancers of the acoustic nerve and of the parotid gland) and to retrospectively evaluate the possible risks to users of mobile phones.

Suspicions feed the debates that focus on the materiality of the observed effects, their description, and the causal chains responsible for them. Only through systematic investigations can these suspicions be invalidated or confirmed. As the exploration of possible states of the world progresses, the controversy may evolve; suspicions may gradually give way to presumptions.

From Suspicion to Presumption

Suspicion leads to the contemplation of states of the world which are considered to be plausible in the light of bizarre, fragile phenomena that are difficult to describe. With presumption we move on to a new stage. In law, the term 'presumption' designates induction from a known to a disputed fact. The corpse exists, and conjectures lead us to think that we have found the murderer, but we do not have the proof that assures us that he or she is the real culprit. In the controversies corresponding to this case, the phenomena are firmly established and no one challenges their existence. Sound observations enable one to back up the facts and qualify them by showing, for example, that thresholds have been crossed and developments confirm the observations: the number of deaths cannot be explained by random phenomena, and their number exceeds levels beyond which the tendency is irreversible. The uncertainties focus essentially on the causal chain, although we have the beginnings of an explanation. Such was the case with Bovine Spongiform Encephalopathy (BSE) in 1988. The threat was certain. We knew that cows were affected by it; we knew what the agent was, but its existence raised some doubts; we did not know exactly how it spread, but some hypotheses seemed likely; we did not know if the disease could affect humans, but nothing could be ruled out. In such situations, controversy essentially focuses on two points. First, as in cases of suspicion, the reliability of the information and the data collected may be disputed. Do they merely reflect the anxieties of those involved in publicizing the problem, or are they the firm basis of a scientific evaluation of the dangers incurred? The confrontation may also, and especially, focus on the action to be taken. Do we know enough to make decisions? Should we undertake further investigation in order to stick with indisputable proofs? If so, what tracks should be followed? Should we wait before taking measures, or should we take them right away? If we opt for the latter, what measures is it appropriate to adopt?

The issue of nuclear waste corresponds quite closely to this scenario. No one denies the dangers of storage; the debate concerns how to deal with them. Should we put up with irreversible storage that some specialists say presents only a low risk? Or should we pursue new lines of research in the hope that they will result in methods that will enable us to eliminate the danger associated with nuclear waste? In the meanwhile, what measures should we take?

Social and Technical Uncertainties

At first sight, the uncertainties we have so far considered could be described as scientific or technical. The strategy that is essential for lessening them could come from laboratories or research departments.

However, the controversies engendered by these uncertainties go far beyond solely technical questions. One of the central things at issue in these controversies is precisely establishing a clear and widely accepted border between what is considered to be unquestionably technical and what is recognized as unquestionably social. The line describing this border constantly fluctuates throughout the controversy. To declare that an issue is technical is effectively to remove it from the influence of public debate; on the other hand, to recognize its social dimension restores its chance of being discussed in political arenas.

Nuclear energy provides, at least in France, good examples of these fluctuations. In the 1960s the issue of nuclear energy was seen as being essentially a technical matter and therefore as having to be dealt with by the relevant specialists; the social was defined in a residual way as rallying a public that was more or less favorable, more or less prey to irrational fears and anxieties. Twenty years later the division had undergone profound change. The anonymous public constituted by the individuals of crude psychology gave way to differentiated groups capable of speaking outside of opinion polls and of developing constructed arguments. It was enriched by genuine political movements that challenged the democratic character of certain decisions. After another ten years, the stage was crowded with unexpected actors: residents' associations, local groups, chicken farmers, viticulturists, professional associations. What the anthropologist Marilyn Strathern calls the "proliferation of the social"[5] was accompanied by a continual enrichment of the technical issue itself. In truth, the two histories are closely interwoven. That is why the initial distinction becomes blurred. To the question "Is deep burial a technical solution?" everyone agreed in giving an affirmative answer. To the question "What is the social component of the nuclear issue?" the specialists answered with a single voice: "It arises from the public's irrational fears." Thirty years later, this response seems out of date. This society without consistency has vanished before the disenchanted eyes of nostalgic technocrats. Multiple groups have appeared whose existence no one suspected, defending their interests and projects, and adding their two cents to the so-called technical discussions. There are, of course, many people who contest the solutions envisaged or who demand their modification. But life is not that simple. Security and surveillance services are also summoned and questioned as to their long-term ability to fulfill their mission; there are even the "future generations" about whom everyone is suddenly concerned, in whose name all believe they are authorized to speak, and who are thus invited to all the meetings at which storage, fast breeder reactors, and transmutation are discussed. As a result, the solution of deep burial is only secondarily seen as technical problem. To the great displeasure of the specialists, it becomes an eminently social and political problem. The border between the two spheres has been completely scrambled in the space of two or three decades.

As the foregoing example shows, the controversies that unfold in hybrid forums are fostered not only by scientific and technical uncertainties but also by social uncertainties. In discussing the border between what is technical and what is social, the protagonists, whose identities vary over time, introduce an indeterminacy that will not be settled until the end of the controversy. Moreover, it is the entry of new actors on the scene that causes the border to be called into question. Society may indeed be as uncertain and unpredictable as the nonhuman entities with which it has chosen to share its destiny.

Dynamic

Socio-technical controversies unfold in time and space. Their trajectory is largely unpredictable because it depends on the nature and degree of the uncertainties and also on the way in which some of them end up being lessened or disappearing. What social groups will arrive on the scene? What alliances will they forge? What technological options will be revealed, or ruled out, by the research undertaken? What new lines of research will be explored? These questions are continuously formulated and reformulated as the socio-technical controversy develops. They are both the consequence and the motor of its dynamic. To understand this point, it is useful to return to the notion of a possible state of the world.

We have said that in a situation of uncertainty the states of the world that are likely to be realized are to a great extent unknown. There is reliable evidence that permits us to think that the list of conceivable scenarios is not exhaustive, that each scenario is only described schematically and very incompletely, and that the causal chains that allow us to predict the conditions under which a scenario can or cannot be realized are only identified approximately. Controversy focuses on these zones of ignorance. It explores them and occasionally helps to reduce them through the game of confrontations to which it gives rise and through the information it generates and circulates. In short, it organizes the more complete investigation of possible states of the world. Thus we pass from radical uncertainty to suspicion, and then from suspicion to presumption and sometimes proof. But this is not the only possible trajectory. Uncertainties may increase with the emergence of increasing numbers of diverse groups and the discovery of vast continents of ignorance.

BSE is a good example of a situation of uncertainty that took a long time to reduce and which is present to some extent even today. Although the epizooty now seems to be under control (1,646 cases in the world in 2003, 878 in 2004, and only 474 in 2005), for many years the course of this "crisis" was characterized by a real proliferation of uncertainties. In the mid 1980s, for example, two main means of transmission of prions were identified: feeding animals with contaminated meal and transmission by affected cows to their calves during gestation. Yet despite culling and strict control of the animal feed sector, the number of cases of cattle with BSE born after the ban remained stable albeit low (16 in France for the first half of 2000). Because the origin of this type of contamination could not be explained via the two known routes, complex hypotheses were put forward. Some of them had already been formulated in 1999 by expert committees, and used by the French government to oppose the lifting of the British beef embargo, despite the European Commission's demand. In particular, the existence of a third contamination route was suspected, but none of the observations made nor the measures taken during the heat of the controversy were able to reduce the uncertainties. Nothing pointed to the outcome of this turbulent controversy, which was constantly fueled by new questions. Rather than reducing uncertainties, the investigations tended to amplify them, especially at first.

One of the powerful motors of this dynamic is found in the dialectic established between scientific and technical research on one side and social reconfiguration on the other: it is decided to undertake investigations that result in the identification of new possible states of the world, mere reference to which brings out unforeseen actors, who, in turn, launch themselves into the debate and propose new lines of exploration. The socio-technical spiral is up and running and has no reason to halt. Given its fruitfulness—it produces knowledge and fosters learning—the only reasons for halting it are bad ones, despite the fears aroused by its development.

Explorations and Collective Learning

Sociologists of social movements have shown how easy it is for social conflicts to be assimilated to pathological forms of behavior that can be explained either by the irrationality of those who are mobilized or by the clumsiness of the dominant actors. Socio-technical controversies are not exceptions to the rule. They are often seen as the result of a lack of communication and information: the scientist or politician did not want (or failed) to be understood by the ordinary citizen. At best, controversies are often seen as a waste of time that could be dispensed with; at worst they are seen as the hardly avoidable consequence of the intellectual backwardness of people in need of continuous guidance.

The position we take in this book is at variance with these two conceptions. It is that controversies enrich democracy.[6] When scientific expertise and political voluntarism adopt the form of an authoritative discourse, they fail to respond to the questions of concerned citizens.

We propose to shift the gaze cast on controversies by passing from the time of contempt or indifference to one in which they are taken into consideration. This is not out of an indiscriminate love of exchanges and communication; as we will show, controversies are not just a useful means for circulating information. Nor are they reducible to simple ideological battles. With the hybrid forums in which they develop, they are powerful apparatuses for exploring and learning about possible worlds.

Controversy as a Mode of Exploration

Controversies make possible the exploration of what we propose to call *overflows* engendered by the development of science and techniques. Overflows are inseparably technical and social, and they give rise to unexpected problems by giving prominence to unforeseen effects. All, specialists included, think they have clearly defined the parameters of the proposed solutions, reckon they have established sound knowledge and know-how, and are convinced they have clearly identified the groups concerned and their expectations. And then disconcerting events occur.

To start with, controversies help to reveal events that were initially isolated and difficult to see, because they bring forward groups that consider themselves involved by the overflows that they help to identify. As investigations go on, links from cause to effect are brought to the fore. The controversy carries out an inventory of the situation that aims less at establishing the truth of the facts than at making the situation intelligible. This inventory focuses first on the groups concerned, on their interests and identities. It is not the result of a cold, distant, and abstract analysis. It is carried out at the same time as the actors arrive on the scene. The distribution is not known in advance but is revealed as the controversy develops, and it is precisely for this reason that the latter is an apparatus of exploration that makes possible the discovery of what and who make up society.

The sudden appearance of new actors (residents living along a polluted river, consumers of beef, pregnant women in the canton of La Hague, future generations who will inherit irreversible stocks of nuclear waste) corresponds to more or less radical reconfigurations of the social landscape. In the first scenario it may be a case of new actors who are not really new. Previously kept in the wings, they take advantage of the controversy to enter the scene in a legitimate role. The second scenario is that of really emergent concerned groups created by the controversy.

The example of the protest in France against the TGV Sud-Est (South-East High-Speed Train) illustrates this dual process perfectly, as in many other countries. To begin with, when the first studies are completed, in July 1989, the Société Nationale des Chemins de fer Français (French National Railways) initiates institutional yet discreet consultation, with the leading politicians only. Subsequently, at the beginning of the 1990s, after leaks about the route and the revelation of the existence of these contacts, there is an outburst of mobilization. Elected representatives from the communes and departments, associations for the protection of the environment, representatives of wine growers and market gardeners, and, in some areas, a number of residents associations, all come together in a heterogeneous coalition. This proliferation of actors and demands halts the project and results in the postponement of the start of work. An arbitration mission is appointed in August 1991 to offer the threatened populations "a bunch of new negotiators."[7] But this remedial operation, which lasts until the start of 1991, is not enough to reduce the conflict. In parallel, actors from local politics and associations form a structure. A local association mixing farmers and residents is formed at the start of 1990 and leads protests that produce a more entrenched situation. Shortly thereafter, a more extensive coordination is created and brings together very diverse groups on the theme of the defense of Provence's landscape. It initiates a new representation of associations that rivals the older regional organization, which is not very involved in the protest, and it plays a decisive role in the third stage, in the course of which a pluralist "college of experts," appointed in May 1992, conducts the negotiations that lead to the resolution of the crisis two years later. We can see how, in this case, the controversy brings to light actors who previously were distant from the public space or did not exist.

Socio-technical controversies contribute to the realization of a second inventory: an inventory of the possible connections between the problems under discussion and other problems with which some committed groups strive to establish links. The effort to make links is not just a matter of simple exposure. It needs the appearance of new actors and their activity of reflection and investigation to establish unexpected connections. Decision makers think that the parameters of the questions to be dealt with have been suitably and properly defined, from both a technical and a political point of view, and now overflows identified by the actors demonstrate the opposite: that controversy allows an inventory to be made of the different dimensions of what is at stake in a project. Controversy brings about the discovery, for example, that the mobilizations provoked by the introduction of major facilities (motorways, high-speed trains, airports, or the storage of dangerous waste) is not explained simply by the fear of pollution experienced by the resident populations, but also by their relationships with the territory, its history, and its elites.

We can say that the controversy enriches the meaning of a situation. In fact, all big projects of development or social reform pursue precise but partial objectives. They generally respond to needs or demands which are deemed to be legitimate and which come from a public agency or body seeking to extend or renew its field of action (modernization of the means transport, resolution of the problem of nuclear waste, or even broadening anti-drug policies); they may also arise from political parties seeking to deal with problems encountered by the population (new epidemics, lack of security, the lack of status for civil partnership, etc.). The initial delimitation and formulation of these needs is generally carried out within closed circles (political offices, central administrations, directors of public enterprises, and so on). But such containment cannot last. Every decision-making process requires a work of opening out, of diffusion, if only because of the need to mobilize the actors who will enable the project to be brought to a successful conclusion (or, at least, will guarantee that it is not violently rejected). Deciding is opening Pandora's Box by permitting actors previously held at arm's length to take part in a dynamic to which they quickly contribute.

The development of mobile telephony perfectly illustrates this open process of exploration of issues and matters of concern. When the first relay antennas were set up, nobody took any notice. But information soon began to circulate. Researchers claimed that the electromagnetic waves emitted by the antennas could affect the health of people living nearby. Local organizations were set up and demanded that the plan to install the antennas be shelved. International epidemiological investigations were launched and produced results that were reassuring but left many doubts. The health issue continued to be a subject of mobilization, and many measures were taken, at European and national level, to set emission levels. The experts kept on working and writing reports. At their suggestion, the French government, inspired by the precautionary principle, decided to go further and demanded that antennas not be installed near nurseries or schools. But soon things became complicated. The health issue became only one among other controversial issues. People who lived near antennas and who had started by questioning their placement in the name of health often switched to other subjects of preoccupation. For instance, they denounced the conditions under which the local authorities had decided to install the antennas, or they criticized its poor environmental integration. On a site on which unexplained cases of leukemia appeared, families started by implicating the antenna, placed on a school building. One thing led to the next, as official and unofficial inquiries proliferated. It was discovered that the ground had been polluted by a military camp situated nearby, and by industrial waste. Thus, the history of an entire area was examined by the population, and health concerns were soon forgotten. The people living in the area laid charges against the municipality, which it accused of having chosen the site without any public consultation, and against the mobile phone operators who devalued public property by installing antennas that defaced the buildings. In short, at national and local level we witnessed an ongoing exploration of matters of concern. These proliferated and ended up weaving a dense web of unexpected issues and groups expressing and exploring them.

These stories and other examples in this book illustrate the power of socio-technical controversies to reveal the multiplicity of stakes associated with one issue, but also to make the network of problems it raises both visible and debatable.

Controversies also allow the exploration of conceivable options by going beyond the list established by the official actors. Thus the public debates provoked by certain bullet train projects succeeded in reopening the "black box" of technical solutions. While the TGV no longer gave rise to discussion after the success of the Paris-Lyon link, which was thought to be not only the best solution possible but the only conceivable solution, on the occasion of the TGV Sud-Est project it was possible to reintroduce another option: that of the tilting technique, which had initially been rejected. In a situation of a lack of public funds, the mobilization of new political actors (local communities, groups defending the environment, and residents associations), and the development of controversies over all TGV projects, this alternative solution was re-launched and even became popular. Certainly, the tilting train is defended only by minority groups and is firmly criticized by the Société Nationale des Chemins de fer Français. But it becomes an obligatory subject of debate in public exchanges. Everyone taking part in the debate is now required to make their position public and to argue for it.

A controversy reveals uncertainties and, as a consequence, new lines of research to be explored. It provides the opportunity to return to abandoned tracks, for one of the strategies for re-opening a debate or for changing its terms is to mobilize solutions that have greater credibility, having already been tested in other places and other circumstances. Faced with realistic options that they did not think they would have to consider, those promoting a project have to justify themselves, explain why they do not want to, and thereby make explicit the criteria for their choices and decisions. By situating a policy in its history, or by redefining its context, controversies bring to light possibilities that were not taken up and suggest the recycling of solutions envisaged in the past. In addition, they lead to the identification of constraints that were not taken into account during the development of technological projects. Once identified, these new constraints will reorient research and open up the elaboration of new projects and new solutions.

Because they formulate a triple inventory of actors, problems, and solutions, controversies are a highly effective apparatus for the exploration of possible states of the world when these states are unknown, owing to uncertainties. They encourage the enrichment and transformation of the initial projects and stakes, simultaneously permitting the reformulation of problems, the discussion of technical options, and, more broadly, the redefinition of the objectives pursued. This exploration, which aims to take the measure of overflows not yet framed within definite parameters, equally constitutes a process of collective learning.

Controversy as Learning
Once the overflows are brought out and made explicit, the question is no longer whether or not a solution is good; it is a question of how to integrate the different dimensions of the debate in order to arrive at a "robust" solution. The opposition between experts and laypersons, between science and politics, is replaced by socio-technical arguments, by scenarios that articulate different kinds of considerations. Conflict is not extinguished, but shifted. Controversy allows the design and testing of projects and solutions that integrate a plurality of points of view, demands, and expectations. This "taking into account," which takes place through negotiations and successive compromises, unleashes a process of learning. This learning is not limited to redrafting the proposals of experts, who could then be content with integrating non-technical considerations so as to take them over. In some extreme cases, such redrafting takes the form of a simple modification of vocabulary in order to avoid words that frighten the population. Since the 1991 French law on nuclear waste, we no longer talk of "burial," but of "deep storage." Talk of creating an "underground laboratory" defers the debate on the creation of storage centers. The learning pro-

voked by socio-technical controversies goes further. It is collective. As the following chapters will show, it allows laypersons to enter into the scientific and technical content of projects in order to propose solutions, and it leads the promoters to redefine their projects and to explore new lines of research able to integrate demands they had never considered.

To what are these effects of learning due? First, to the constraints that every organized debate in a public space brings to bear on the actors involved. In the dynamic of controversy, everyone is asked to listen to other people, to respond clearly to their arguments, and to formulate counter-proposals. A "besieged fortress" type of strategy (defending one's initial point of view at any cost), or one of "sitting on the fence" (saying as little as possible to avoid committing oneself), is especially unproductive, and generally such strategies go against those who adopt them. In a public arena, the actors must express themselves and listen. This double requirement results in real exchanges taking place.

But exchanges alone are not enough, however courteous and civilized. A gain must be produced. New knowledge must be acquired and shared, and new ways of thinking, seeing, and acting must be developed, pooled, and made available. Two fundamental mechanisms account for the production of this gain.

The first mechanism is linked to the unusual confrontation that socio-technical controversies organize between specialists and laypersons. Controversy establishes a brutal short circuit between these two poles, which are usually separated by an almost unbridgeable gulf. In fact, relations between specialists and non-specialists usually bear the stamp of asymmetry. The former, imagining that they are faced with an ignorant or even obtuse public, take on the mission of enlightening and instructing the latter. The discussion established in hybrid forums wrong foots this model. It demonstrates that both categories of actors possess specific forms of knowledge (a capacity for diagnosis, an interpretation of the facts, a range of solutions) that mutually enrich each other. In the case of the TGV Sud-Est, the residents unfavorable to the project give prominence to new local problems (the construction of massive embankments, the environmental impact on sensitive natural milieus, unawareness of local transport networks) which were not considered in the initial studies and with which the experts have to make themselves familiar and which they will have to learn to take into account. In the Rhine-Rhone TGV project, the laypersons also help to put the experts in a learning situation. The arguments of the opponents marshal facts that had already been collected by groups opposed to a previous project for a canal with the same course, and which the promoters had not explicitly taken into account (in particular regarding the impact on the hydrological network).

The second mechanism of learning is linked to the perceptions that different groups have of each other. Instead of confronting each other and debating through interposed spokespersons and official representatives (members of parliament, local councilors, union leaders, et al.), the actors involved in the controversy do not hesitate to provide themselves with new representatives closer to their way of thinking and demands. The latter, having no guarantees that they will keep their position (they can be disowned at any moment), take better account, in the positions they adopt, of the evolution of changing and developing identities. The actors involved find themselves more directly in tune with each other, which improves mutual understanding. A socio-technical controversy makes it tangible that planners are not just developers, that opponents of nuclear power are not just nostalgic for candlelight, that the councilors of small communes are not just simple spokespersons for their electors, and that scientific experts are not just monsters of abstraction indifferent to any social cause. Controversy makes it possible to go beyond a simple opposition setting defenders of the general interest against defenders of selfish interests, or representatives of progress against the standard bearers of a backward-looking mode of life. For a time, the relative equalization of "rights to speak," the opportunity for everyone to argue on his or her own account and to question the justifications of others, transforms for a time the usual hierarchies and their underlying conceptions. This mutual discovery obviously affects each actor, whose identity is modified in turn. Becoming aware that one's sworn enemy is not the person one thought he was facilitates the revision of one's own positions.

The redefinition of identities opens the way to compromises and alliances that would be unthinkable without the existence of controversies. The latter thus contributes to the formation of networks of actors sharing a collective project, to the emergence of "project" or "cause" coalitions that otherwise would not have existed. These reconfigurations of identities, proximities, alliances, and commitments result in a veritable mutual learning process that is all the more fruitful as the traditional representative institutions are powerfully short-circuited. Controversies make it possible to overcome the gap separating laypersons and specialists, but also to go beyond the sterile roles of the ordinary citizen and his legitimate representatives that tend to prevail.

The Dialogical Space of Hybrid Forums

The examination of the functioning of hybrid forums leads us to see the controversies that develop within them as powerful and original apparatuses for exploration and learning:

• exploration of the identity of the actors who are concerned about the projects under discussion; exploration of the problems raised as well as all those that the concerned actors consider to be associated problems; exploration of the universe of conceivable options and the solutions to which they lead

• learning that results in alternate exchanges between the forms of knowledge of specialists and the knowledge of laypersons; learning that, beyond institutionalized representations, leads to the discovery of mutual, developing, and malleable identities that are led to take each other into account and thereby transform themselves.

Controversies are not summed up in the simple addition and aggregation of individual points of view; their content is not mechanically determined by the context in which they unfold; they are not confined to friendly discussions or by debates intended to conclude with an agreement. By trial and error and progressive reconfigurations of problems and identities, socio-technical controversies tend to bring about a common world that is not just habitable but also livable and living, not closed on itself but open to new explorations and learning processes. What is at stake for the actors is not just expressing oneself or exchanging ideas, or even making compromises; it is not only reacting, but constructing.

By fostering the unfolding of these explorations and learning processes, hybrid forums take part in a challenge, a partial challenge at least, to the two great typical divisions of our Western societies: the division that separates specialists and laypersons and the division that distances ordinary citizens from their institutional representatives. These distinctions, and the asymmetries they entail, are scrambled in hybrid forums. Laypersons dare to intervene in technical questions; citizens regroup in order to work out and express new identities, abandoning their usual spokespersons. Thanks to this double transgression, as yet unidentified overflows are revealed and made manageable. The hybrid forums could thus become an apparatus of elucidation. The cost of accepting their use is acceptance of the challenge to the two great divisions. Actors involved in socio-technical controversies are not mistaken. When they establish a new hybrid form, they lay their cards on the table: "We do not accept the monopoly of experts! We want to be directly involved in the political debate on questions that our representatives either ignore or deal with without speaking with us!"

Every hybrid forum is a new work site. It is a site for testing out forms of organization and procedure intended to facilitate cooperation between specialists and laypersons, but also for giving visibility and audibility to emergent groups that lack official spokespersons. The task of the actors is all the more difficult as it comes up against two monopolies: that of the production of scientific knowledge and technology and that of political representation. Without a minimum of formalism and guarantees, hybrid forums would be doomed to failure, a protest soon to be forgotten. By designating the great double division as that which they are struggling against, the actors express this clearly. They identify the possible adversaries; they get ready for a confrontation. This would quickly redound to their disadvantage if there were not procedures that the actors had invented and tried out, forum after forum. Chapters 4 and 5 present these procedures and put forward a balance sheet of the experience so far. But before doing this we must examine the question at the heart of technical democracy: In what circumstances, under what conditions, according to what modalities, and with what effectiveness is collaboration between laypersons and specialists conceivable? Is it not, perhaps, just a case of occasional and superficial exchanges? Alternatively, can we conceive of a lasting cooperation? This is the theme of the next two chapters.

2   Secluded Research

The following recollections are from Callon's notebook.

I no longer recall when I met Emmanuel for the first time. While waiting with my son for the observatory service elevator, I remember that only a few weeks earlier we were together at Béziers. The local astronomy society had organized a meeting between amateur and professional astronomers. Emmanuel, who had struggled throughout his career to encourage these meetings, had invited me. Everything had begun, around 8 p.m., with a Pantagruelian cassoulet washed down with plenty of wine. Then, before getting down to the heart of the matter, we were dragged to see the exhibition of books where works, each more learned than the one before, and a good half of which were written by non-specialists, were on display. We finally gathered in a large room. The amateur astronomers from all over the South of France had invited some professionals to give some lectures; they had even organized a round-table discussion and asked me, as a sociologist of science, to talk to them about the role of amateurs in the history of their discipline. On the stroke of midnight, Emmanuel launched into his talk. At 1 a.m., tiredness playing its part, we found ourselves in a fierce discussion, backed up with mathematical formulas, about the possible existence of life on other planets.

Astronomy is a science in which amateurs and professional have always organized fruitful exchanges, and they continue to do so. "Come to the observatory of the Pic du Midi," Emmanuel had said to me, "you will see that there is active collaboration and it doesn't cause any problems."

I had carefully followed the directions on the fax Emmanuel had sent me when he learned that he was entitled to a week of observation at the end of the summer. "At Tarbes, you take the road for Bagnères, and from Bagnères you go to La Mongie and then to the Tourmalet pass. At the pass you will see a road on the right for the peak, which will take you to 200 meters from the summit. There is then a steep 20-minute climb on foot. You arrive at the observatory by the North terrace and ask the guides or the person in the refreshment room to tell me you have arrived. As a rule I will be at the 2-meter telescope, or else on the terrace if the weather is fine."

The service elevator has a wooden floor and runs on a rail equipped with a rack. Along the winding path that unfolds beneath our feet, the last tourists are going back down, preceded by their shadows that stretch out in the falling light. The astronomers are taking back possession of the observatory. They swarm over the terraces in order to take advantage of the last of the sun's rays before the trying night that awaits them.

The site is magnificent. The observatory is set delicately above the plain and dominates the neighboring peaks. In whichever direction you look, there is no obstacle.

Emmanuel is waiting for us on the arrival platform. We cross a few dozen meters casting a last glance at the tawny vultures lazily sweeping on warm currents of air in wide circles up toward the sun.

The door is closed on us. The feeling of having climbed aboard a ship of the open sea, as well as of having hardily won the right to be overwhelmed by light and the bracing air vanishes brutally. We enter a labyrinth of dark galleries; we are plunged into the confined atmosphere of a submarine.

The observatory was designed to allow for a cloistered life, cut off from the rest of the world. We realize that the long route we have followed was not intended to transport us to open spaces, but to distance us from them. Furthermore, when the snow blocks the few windows situated at the levels that have not been cut into the rock, the observatory is transformed into a huge black chamber, a subterranean town, in which one can move around without putting one's nose outside. Thus turned in on itself, the observatory resembles a Cistercian abbey that was not designed for the contemplation of God, but for the equally silent contemplation of the planets and stars. Apart from the telephone lines, the service elevator, and the cable car, which works only irregularly, the only link with the outside is that provided by the three telescopes allowing observation of the sky.

Emmanuel works with the 2-meter telescope. This telescope was named Bernard-Lyot, in homage to man who, in 1930, carried out the first observations of the solar corona with the help of a coronagraph, an instrument he invented that enables total eclipses to be produced at will by blocking out the heart of the sun. Emmanuel is tracking the twin galaxies in order to understand how they are born and how they die. We join him just as the sun is setting. He is delighted with the weather, for it promises to be a clear night. It should give him long hours of high-quality observations.

Observing the sky through a telescope! It is an unfortunate expression. Emmanuel is installed in a room full of screens. The telescope is invisible and you might even doubt its existence if it were not for the dreadful creaking that breaks the silence of the night when the technicians maneuver it by remote control in order to direct it at new celestial bodies. These cannot be observed directly. The signals they emit are filtered, calculated, and worked on before being displayed on computer screens. The colors and contrasts are not natural; they are obtained by means of complicated algorithms. There are no less than nine screens in the room. Some are to control the dome, others are used to point the telescope at the part of the sky where the galaxies being studied are found, others enable the image to be worked on, and its contrasts or legibility to be modified, and others, finally, give access to the data banks available on the Internet that allow continual comparisons and verifications to be carried out. Not even the television screen is missing, on which, between adjustments, the technicians follow an American series.

The room is so cut off from the world, so immersed in what some would describe as virtual realities, that at times amusing coincidences take place. One of the technicians manages to fix the reference star on his screen. By putting the luminous spot of the star in a small white circle, he will be able to delegate to the star the task of keeping the telescope in the right direction. "That's it, got it!" he says to those around him. Seconds later, a hired killer on the TV channel France 2, having followed his target in the sights of his rifle and dispatched him to the next world, says "Touché!" Each has fulfilled his contract to bring two images together—one to control the movements of the telescope, the other to execute his victim.

A little later, Emmanuel draws me from my reveries: "Michel, can you go and look at the sky?" It is 5 a.m. and I am beginning to doze off. I go through the little office alongside the room of screens, stand on a footstool, and open the tiny window. The fresh air wakes me up. Emmanuel approaches me: "Do you see any clouds?" Not noticing the comical character of this question from a professional astronomer installed in a high-tech observatory, I answer mechanically: "No, the sky is completely clear." Emmanuel returns to his screens. I hear him say to the young student who is helping him: "It is an artifact; Michel has told me that there are no clouds."

The 2-meter telescope of the Pic du Midi observatory is so secluded, so cut off from the world, that it only communicates with the outside world

through interposed screens. It clearly draws all its strength from this seclusion; the site permits observations that are more direct, less disturbed, and
less blurred by external interference than those of the naked eye or of an
eye stuck to the eyepiece of an astronomical telescope. Imaging techniques
involve maximum elimination of human intervention and the bias they
might introduce into the tasks of observation.

However, to the specialist, this seclusion still seems insufficient. Hence
Emmanuel's strange question, which he addresses to me, a poor sociologist
who knows nothing of the sciences of the universe! The Pic du Midi observatory is still too much in the world, too dependent on it, since simple
clouds can impair the accuracy of its instruments and disrupt the observations. Because it is not secluded enough, not sufficiently independent of
the surrounding world and its sometimes tiresome contingences, the Parisian decision makers, inspired by ultramodern astronomers, have it in
mind to dismantle it. And yet, when I arrived, it seemed to merge so well
with the rock on which it was built and with which it was joined, that I
had the feeling that nothing threatened it and that it was part of the
mountain. Outmoded! Such is the specialists' irrevocable diagnosis. This
observatory, whose history goes back to the previous century, is outmoded.
It has been rendered outmoded by the Hawaii observatory, and by satellite
telescopes. The purity of the Pyrenean sky, the altitude of the Pic du Midi,
and the isolation of the site are no longer sufficient advantages. Even here,
sheltered from everyone and everything, the observatory is still too
attached to the surrounding world. The progress of knowledge forces its
seclusion, and the distance it takes from the ordinary world, to be pushed
further. At Hawaii, the clouds are less of a nuisance, the air is even purer,
and the conditions of observation are even better. But the break between
the laboratory and the world obviously reaches its peak when the observatory itself is loaded on a satellite. In that improbable spot, far from human
beings, we can get as near as possible to producing a truth that no pertur-
bation will veil.

The history of astronomy, like that of many other sciences, is one of the
pursuit of an extreme seclusion. One of the ideals of Western science seems
to be to establish its laboratories and install its instruments not only as far
as possible from the world in which we live, but also out of reach of amateurs and laypersons. Nowhere is this uprooting from the world more spectacular than in the case of astronomy, a discipline that has lived and
survived thanks to the inquisitive enthusiasm of people who have not chosen to make it their profession. The history of the Pic du Midi is in fact one
of long and fruitful collaboration between specialists and amateurs. Here

more than elsewhere, they work together, take their meals together, and share the same instruments and observations. On some subjects, like the observation of the solar system and its planets, amateurs have even acquired an international reputation. And now, in the name of the pursuit of seclusion, amateurs will no longer be able to work with professional astronomers. The Pic du Midi is not high enough, it is not isolated enough. In future observatories, no place is foreseen for amateurs, and even less for laypersons like my son and myself who dream of a simple visit. The places where knowledge is produced constantly become more remote and out of reach of whoever does not belong to the inner circle. Undoubtedly, the divide is getting wider between those who have the right to answer the questions they ask because they have access to the instruments, and those who only have a right to frozen knowledge.

Of course, complete exclusion will not be possible in the case of the Pic du Midi. The place is too charged with symbols for the crime to go unnoticed. Resistance was organized. After a long struggle to save the site, amateurs have finally been authorized to use the two telescopes. Secluded science, magnanimous out of necessity, allows non-specialists to use instruments for which it no longer has any use. Despite this arrangement, the divide is truly consummated. The Pic du Midi observatory is no longer on the map of professional research. It needed time, a lot of money, great determination, and hard-heartedness. But nothing and no one could resist science on the move. It will no longer be possible to enter future observatories by a service elevator, because they will have been put out of everybody's reach. ∎

The pursuit of seclusion affects every area of scientific research, sparing no discipline. Particle physics shuts itself away and buries itself in ever more powerful accelerators; biology is not slow to follow; trying to decode different genomes, it becomes burdened with increasingly effective sequencers. Even the social sciences, following the example of economics, share this destiny. Why this obsession with seclusion? What are its benefits for research? What difficulties does it create?

Before analyzing the mechanisms of this seclusion in detail, it is worth trying to understand how we arrived at this great division between laypersons and scientists so as to elucidate the reasons for its effectiveness, but also to identify better the problems it raises. In a word, we need to trace, albeit briefly, the history of the gradual establishment of what can be called laboratory research, of research that has distanced itself from the world in order to increase its productivity. After this detour we will be able to conceive how, at what price, but also with what advantages, the links can be renewed and the bond restored between those whose profession is to produce knowledge and those to whom this knowledge is immediately or distantly addressed.

The Great Confinement

The start of a new millennium is always a difficult moment. Western societies are threatened by two demons that pull them in opposite directions. The first, certainly a bit aged, but still very much alive, sank us and our forebears in a blind belief in scientific progress. Science, it asserted at the top of its voice, is the best guide for leading humanity, if not to abundance, then at least to affluence, while protecting us from all kinds of obscurantism. But no sooner had we begun to doubt its word, no sooner had we unmasked it, than it reappeared in new clothes. Reading on our face a disappointment as deep as our hopes had been high, it does not hesitate to burn the idols it had got us to worship. Certain of the effect, it cries out "God is dead, progress is dead." It sees us hurt and urges us on to nihilism and absolute relativism. "The dream of scientism is only a nightmare. Science brings with it corruption, as clouds bring the storm." And how can we not believe this? In our head we all have the disastrous chain reaction: Einstein dreamed the bomb, Roosevelt decided to build it, Oppenheimer constructed it, and Truman ordered it dropped. Nuclear physicists have introduced into our world these strange atoms that disintegrate; the pilot of *Enola Gay* gave them their freedom one day in August 1945. We know the result of thousands dead, and the Hiroshima museum where pacifists from all over the world gather in memory of the people irradiated; but Oppenheimer too (who groans in Truman's office "I have blood on my hands") and Truman ("Don't bring me that accursed madman any more. He didn't launch the bomb. I did."). The second demon, avatar of the first, sniggers; he knows that we are ready to come to the conclusion that science and technology are social through and through, the consequences of the will to power. Just as he knows that he will have no difficulty convincing a handful of despairing philosophers and sociologists to proclaim *urbi et orbi* that reason is dead, murdered by unreasonable beings who thought themselves reasonable, too reasonable.

The demons have triumphed. They force us to choose between the plague and cholera, between science established as an absolute, as neutral (its benefits or damaging effects depend only on our will), and science corrupted by power or the prisoner of its own cultural *a priori*. To avoid this trap and escape from our devils, we must retrace our steps to find the mistake that was made. How have we arrived at this strange situation? After having delegated to specialists (researchers and engineers) the task of producing the knowledge and machines that they tell us we need, we can now only think "Should we have blind confidence in them, or should we systematically mistrust what they say and do?" Shut away in their laboratories, they are so distant from us that they have become veritable strangers. And this is why we fear them, especially when they come back loaded with results. *Timeo Danaos et dona ferentes.* They too seem to suffer from this estrangement; they are homesick, they become emotional when they stray from their laboratory benches (like Oppenheimer, who will always regret his success). What is the origin of this seclusion that is the source of so much misunderstanding, resentment, and anxiety?

There are several ways of recounting this history, or rather these histories. We will follow the one set out by Christian Licoppe, because it stays closest to scientific practices while linking them to the social milieu in which they develop.[1] Licoppe proposes to distinguish three great periods in the forms of organization of the production of scientific truth. A different form of laboratory corresponds to each of these stages. Each stage is a step on the road that leads to seclusion.

The Regime of Curiosity

What Licoppe calls "the regime of curiosity" spreads in the seventeenth century. Scientific facts are established in a spectacular manner in the public sphere before an audience of persons whose status renders their testimony credible and trustworthy. This regime is based on the performance of incredible, surprising experiments (*expériences*) that strike the imagination with their unexpected and extraordinary character. They are also based on the existence and presence of this public of distinguished persons, full of aristocratic civility, whose high rank makes it difficult to question their word. This sometimes produces amusing situations. When a comet crossed the skies in 1684, tracked by all of Europe's astronomers, a conflict arose between the observations of the Gdansk astronomer Hevelius and the French astronomers Auzout and Petit concerning the position to be assigned to the comet. This is particularly embarrassing since none of those involved retracts and the competence of all of them is highly esteemed in the community of astronomers. Characteristically, the compromise proposed by the latter allows everyone to save face, at the cost of a small proliferation: There must be two comets rather than one! Etiquette is no less stubborn than the facts.

This regime of an open science, inscribed in the networks of the lettered and the aristocratic, breaks with the still all-powerful Aristotelian philosophy. For the latter, true science (*scientia*) can only be based on empirical statements of common sense, that is to say, a sense shared by all, or be the result of a series of inferences that everyone judges to be true, as in mathematics. Particular facts—those spectacular, original, unexpected phenomena, actually a bit monstrous and departing from common sense—can only form weak and unconvincing links in the demonstrations that deploy them. The learned are therefore reticent with respect to these experimental manipulations, because they are artificial. They produce novel facts for a public whose most serious defect is that it is necessarily a restricted public of a few people.

The position of the new philosophers (the name then given to those who will later be called scientists) comes into direct conflict with that of the old philosophers, since their one obsession is to organize improbable experiments in order to produce phenomena which have never been seen before. The language of the time reveals this opposition by distinguishing *experientia*—which designates the common experience shared by all, including the person we would now call the man in the street—and the *experimentum*—the singular, original experiment, accessible only to the small number of those who have been invited to witness its organization. By definition, *experientia* does not need to be produced in public, since it is coextensive with the public; the *experimentum*, on the other hand, since it is singular and local, confronts the problem of its publicity and the credit it is to be given. The new fact is therefore seen as a spectacle that takes place before a learned and noble audience. That which is constitutive of the truth is that which is put on view to be seen. Reproducibility is a criterion of relative importance. If it comes to it, even the fact that the phenomena may vary has value, enabling witness to play a full role: It is not the *facts* that circulate, but the *accounts* of the monstrosities that have been shown.

The Regime of Utility

According to Licoppe, at the end of the seventeenth century a new regime appears in which new facts are validated in the name of their utility. The reproducibility of "experimentations"[2] and the possibility of de-localizing the instruments used to produce them become central. The scientist makes every effort to make available to those interested, and especially to his reader, all the elements necessary for replicating the experimentation and the effects it produces. Newton, for example, when reporting his optical

experiments, emphasizes that every sensible reader will be able to reproduce the phenomena he describes, based on the separation of the basic colors, with the prisms he employs. Hence, the experimenter's skill becomes
important in this regime. Moreover, in the first half of the eighteenth century the utilitarian demand is expressed in a systematic effort to construct
comparable instruments. Measurement becomes a strategic matter. One
must have calibrated instruments. Réaumur puts this in a striking way:
"Not only do we not understand the language of different thermometers,
each only vaguely understands his own." Hence his obsession: to make
thermometers comparable, defining reference points, like that of boiling
water, "which, being the same everywhere, serves as a fixed point." The
great scientific expeditions assemble travelers who cross continents to carry
out measurements, cramming their trunks with calibrated instruments.
Thanks to them, observations and measurements are taken on the ground
that can be compared, accumulated, and calculated at a particular spot, no
matter where. For example, these measurements enable the length of the
terrestrial meridian to be determined in Paris. The importance accorded to
reproducibility leads straight to theory. The regime of curiosity aimed at
the construction and validation of "isolated" facts. For the regime of utility,
the multiplication of stable, reproducible, and controllable facts enables
one to go back to principles. In 1740, the abbot Nollet sums up the logic
of his work devoted to electricity in this way: "Attentive to the facts, working to multiply them, and carefully reflecting on their circumstances, I
waited more than ten years for them to lead me to the principle from
which they derive; I finally think I can make out this principle and for several years I have been occupied with reconciling it with experience." The
prediction of general principles connected to each other and giving form
to a theory opens the way to new forms of confidence and of the circulation of truth. Belidor gives a precise formulation of this: "Although the
principles I have established are very obvious, they will no doubt be
received with more confidence if I show that experiments on the strength
of wood are in perfect agreement with the theory." A chain of instruments,
disciplined bodies (those of the experimenters), the statement of general
principles, and the formulation of theoretical systems—this is what scientific practices are made of, and what ensures the validity of the results.

The Regime of Exactness
The end of the eighteenth century sees the emergence of the regime of exactness. In France this manifests itself in the requirement to show that the
measurements agree as precisely as possible with the simple and universal

laws reconstructed by theory, which requires the manufacture of increasingly sensitive instruments of measurement. One of the consequences of
this pursuit of instrumental power and precision is the kind of phobia regarding "interference" that takes hold of all scientists. The bodies of the
experimenters and their assistants, and even those of members of the public, are in fact likely to disturb the instruments, particularly when these
instruments are becoming increasingly sensitive. Apparatuses are then confined in laboratories and devices sheltered behind screens, like the thermometers calibrated by Lavoisier, which he protects in a bain-marie and a
double enclosure. Coulomb's balance is so sensitive that the public's presence irremediably disrupts it, so that for experimentations to succeed they
must be conducted in non-public (that is to say, private) spaces. When
Coulomb buries his instruments beneath the Observatory in order to escape laypersons, the last of the Cassini dynasty expresses marvelously the
necessity for this secluded research: "I blocked up and walled off in advance
all the avenues that lead to this spot, except for one reserved for entering it,
but which was closed by a door; in this way I procured for myself a subterranean study with an enormous wall in which, in silence and the greatest
isolation, I was able to pursue these observations, only ever going into the
study alone." This says it all, and in marvelous language. The modern figure of secluded research, withdrawn, cut off from the world, and consequently precise and effective, is born at the same time as its necessity is
explained and justified. The great confinement of researchers has begun.
Doors and windows are closed; we end up together, that is to say, with disciplined researchers and technicians, surrounded by powerful and calibrated instruments. Far from the public and its frippery, specialists form
themselves into communities within which technical discussions can take
place. They are protected from the chatter of laypersons who do not know
what they are talking about, and who cannot know what they are talking
about, the unfortunates, because they are deprived of these laboratories,
cut off from the world, without which no scientific knowledge worth the
name can be produced. The break has never been so sharp. It is summed
up in a series of qualifications that describe science: purity, precision, exactness, distance. This irresistible evolution will be carried to its conclusion by
decades of the Cold War, in the course of which the alliance between scientists and the military will transform seclusion into the isolation of the ivory
tower.

This history, crossed in big strides, is interesting in more than one respect. In the first place, it emphasizes that secluded research, that in which
specialists organize complicated experiments (*manipulations*) with the help

of precise, powerful, and calibrated instruments, is only one possible form of the organization of research, one stage in a historical process that until now has seen at least three. The first corresponds to the first, fundamental break between *experientia* and *experimentum*, common experience and laboratory experimentation. In the regime of the *experimentum*, of experimentation, the essential thing is to succeed in producing the extraordinary, the singular, the not-seen, or the unheard of, in a way that breaks with the routine of *experientia*. Would not the point of departure of all scientific reasoning be this decisive action by which a problem is shown, by which a questioning, an enigma, an oddity is rendered visible, perceptible? Formulating problems, that is to say, following etymology, setting an unexpected obstacle on paths taken a thousand times, is the obligatory passage point of every scientific enterprise.

The second stage is making what was initially singular and local reproducible in different places, and, to achieve this, constructing metrological networks that calibrate the instruments so as to then compare measurements and sometimes arrive at a general principle. Thus, on the way, a community of specialists is formed sharing the same techniques, the same embodied knowledge, capable of comparing and evaluating their experimentations and of capitalizing the results obtained. Then comes the third stage, and with it the time of seclusion, which becomes an obsession. Researchers establish their general quarters in secluded laboratories, sheltered from the public, in order to conduct purified experimentations in complete tranquility, without running the risk of being disturbed by importunate interferences that impede their pursuit of always greater power and exactness. All that remains is for them subsequently to leave their laboratories to present their results and show that their distant exile has not been sterile.

There are three stages then: (1) problematizing (that is, breaking with common experience by making novel phenomena perceptible, and, in order to do this, summoning a public excited by the novelty); (2) constructing a research group that shares the same instruments and is capable of reproducing the phenomena on which they work; (3) cutting oneself off from the world, shutting oneself away in laboratories in order to get to the bottom of things and return to the world stronger. Does not history open up to us, through freeze frames, what the continuous flux of research shuffles and mixes so well that only the final result—secluded research—is visible? Christian Licoppe invites us to discover the ceaseless movements, the permanent exchanges between specialists and the world that surrounds them. By laying out that which has been enfolded, history makes us see that the laboratory is only one element in a larger set-up, one stage in a long succession of comings and goings. It therefore suggests that we dismiss the two sniggering and grimacing devils without choosing between them. Science is no more independent of the will to power than it is its obedient slave. These two illusions are sustained by the image of a science that would be estranged once and for all from the world and its turmoil. If laboratories have distanced themselves, they nevertheless continue to exist within networks of exchanges and interdependencies whose traces the genealogy of secluded research helps us to rediscover.

Translations

One and the same operation of translation enables us to follow the formation and operation of these networks.[3] It comprises three stages. The first is that of the reduction of the big world (the macrocosm) to the small world (the microcosm) of the laboratory. The second stage is that of the formation and setting to work of a restricted research group that, relying on a strong concentration of instruments and abilities, devises and explores simplified objects. The third stage is that of the always perilous return to the big world: Will the knowledge and machines produced in the confined space of the laboratory be able to survive and live in this world? By following these successive translations we will be able to understand the strengths and weaknesses of secluded research.

From the Macrocosm to the Microcosm: *Translation 1*

It would be absurd to leave the public in order to detach oneself from it and then bury oneself in laboratories cut off from the world if something were not preserved in the course of this movement that enables one to turn back to the world with something extra that makes the difference. There is no point leaving for the Americas unless one comes back with pockets stuffed with New World gold! There is no point secluding oneself, shutting oneself away, and burying oneself, as did Coulomb and Cassini, unless this is to gain strength, wisdom, or knowledge! In other words, spotting a good problem, equipping oneself with duly calibrated instruments, and protecting oneself from non-specialists will not be sufficient to make scientific knowledge miraculously flow by itself, like pure water from a natural spring. So what accounts for the supposed superiority of the secluded laboratory?

To answer this question we must return to the movement itself and to that series of breaks: first the break between *experientia* and *experimentum*, then the break between singularity and reproducibility, and finally the break between the laboratory and the world. If these were simple and definitive breaks, they would be sterile. To be fruitful, each of them must combine two mechanisms: that of transportation, which explains that all is not lost, and that of transformation, which explains that something is gained. Thus, at the end of this first move, the big world of common experience, the macrocosm we inhabit, has been replaced by the small world, the microcosm of the equipped laboratory. This reduction, this change of scale (as we speak of a small-scale model), is the source of the laboratory's strange power. The source of the tremendous effectiveness of scientific research lies in seizing hold of the macrocosm and simplifying, pruning, and reconfiguring it in order to manipulate it quietly in the laboratory, completely undisturbed.

To arrive at the highly theoretical works of a Kepler or a Newton required several centuries of observations that were scattered at first, and then rendered comparable and able to be accumulated, especially thanks to the appearance of printing. The instruments of celestial observation had to be progressively perfected so that finally, in a single place, the astronomer's study, the macrocosm could be capitalized and handled on a sheet of paper, reducible to some tables very quickly transformed into a system of calculable equations, and then finally into clear geometrical figures, "representing" the trajectories of the planets. Without the meticulous census of families with a child affected by spinal muscular atrophy, a census mobilizing a myriad of family practitioners not too certain of their diagnosis, but also, and especially, volunteer parents grouped around the French muscular dystrophy organization[4]; without the systematic collection of cells and their storage in banks; and without the extraction of DNA and its analysis, it would have been impossible to fix the origin of the disease, to identify the gene responsible, and to conclude with the simple statement "Infantile spinal muscular atrophy is due to a modification of the gene SMN." The macrocosm selected as starting point—the universe and its celestial bodies in one case, suffering human bodies in the other—has been replaced by successive extractions, abstractions, and reductions to a microcosm that represents it: in one case, a sheet of paper covered with mathematical signs and trajectories; in the other, sequences of bases read on a chromatograph.

This mobilization of the world, which, after being reduced, is transported into the laboratory to be subjected to the tests of experimentation, is common to the natural and life sciences, but also to the social sciences. Think of the databases of the Institut National de la Statistique et des Études Économiques, sociological surveys, the collections of the French Natural History Museum, or the expeditions of scholars to Lapland or Peru, organized at

World

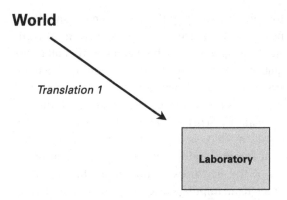

Figure 2.1
Translation 1 (transportation of the complex world into the laboratory).

the height of the Revolution, to collect observations, bring them back to Paris, and calculate the length of the meridian in order to calculate the flattening out of the terrestrial globe. We will call this movement that starts out from the big world in order to arrive at the laboratory, and which replaces a complex and enigmatic reality with a simpler, more manipulable reality, but which nevertheless remains representative *translation 1*. It is translation in the two senses of the word: transport—the world is in the laboratory—and transposition that maintains some equivalences—what is in the laboratory is at once different and similar. (See figure 2.1).

Every discipline, according to its own rhythm and history, passes through the different stages that lead to this substitution, the stages that Christian Licoppe has revealed in the case of physics. There may be a greater or lesser distance between the world and the laboratory, research may be more or less secluded, but in every case there is this detour—*translation 1*—that, if it is well negotiated, assures a certain degree of realistic reduction. Thus researchers may progressively bury themselves in their laboratories, admitting only their colleagues and their instruments. What they study, describe, analyze, and interpret is a purified and simplified world, but, if they have done their work well, it is a world that can be connected up with the big world from which they have taken care to keep their distance.

This is the world translated into the laboratory, reduced to a manageable scale. The relation of force has been reversed. There were thousands of eddies, battalions of cold air colliding with regiments of warm air, marine currents suddenly changing trajectories, and volcanoes interposing their glowing clouds between the sun and the Earth; now, thanks to faithful sensors that every minute send measurements covering the globe, thanks to models loaded on batteries of supercomputers, to mathematical simulations that can integrate hundreds of variables, but also to satellite photographs that transform the atmosphere into a battlefield, we now have meteorological predictions based on printouts churned out by computers, and these predictions cover every point of the globe for a period lasting 48 hours. We had enigmatic, scattered, unrelated diseases with nothing in common, the few doctors with the courage to concern themselves with them accumulating only disjointed observations; now, supplied by the machine, we have a long series of letters in strings on a sheet of paper; those in red are the modifications that explain the disease and transform individual histories into a destiny shared by children whose illness is due to these few modifications. If the secluded laboratory is effective, it is because at the same time as it translates the world it manages a spectacular reversal, an inversion that transforms teeming, dispersed crowds into these traces that can be taken in at a glance. How is this strange takeover brought about? Two notions will be useful to penetrate this mystery: *inscription*[5] and *distributed abilities*. These notions will help us to understand the origin of the incomparable strength of secluded research.

The Research Collective at Work: *Translation 2*
Anyone who has the opportunity to visit a laboratory is struck as much by the proliferation of instruments as by the inscriptions they produce. There are inscriptions that are read on screens, on computer printouts, or in laboratory notebooks in which researchers hourly note the results of their practical activities (*manipulations*). The DNA sequencer and the chromatograph supply images that are strangely like the bar codes that tattoo supermarket goods. The spectrometer draws curves that make one think of the alpine stages of the Tour de France, but where the peaks indicate the possible presence of a substance that has been being tracked for days. The apparatuses of magnetic resonance imaging (MRI) transform the brain into colored maps on which the green or red spots reveal zones of activity. The telescope transmits to the video monitor a digitized portrait of the twin galaxies in the direction to which it was pointed. The detector installed on the particle accelerator supplies data that, when recalculated by sophisticated statistical methods, results in the line of trajectories of collisions, the subtle interpretation of which reveals the brief appearance of a particle whose existence was predicted by a handful of daring theorists. The specimens and samples taken by the zoologist or geologist, carefully classified by their treatment,

transported into the laboratory where they are arranged, compared, analyzed, are then transformed into drawings and integrated into sections, and end up in the form of diagrams supporting the argument developed in an article. The laboratory is a machine for producing inscriptions, for making possible their discussion, interpretation, and mobilization in learned controversies. The famous data (givens) of experience are never given; they are obtained, "made," fabricated. And they take the form of inscriptions that may equally well be photos, maps, graphs, filmed or electronically recorded traces, direct visual observations recorded in a laboratory notebook, diagrams, illustrations, printed samples, 3D models, ultrasound scans, or sonorous spectrums arranged and filtered by techniques enabling them to be visualized. This, and only this, is what the scientist registers, describes, exhibits, analyzes, compares, and measures.

What do researchers gain by taking leave of the "real" world and choosing to concentrate all their attention on fragile traces? Do they not risk giving up the bird in the hand for the bird in the bush? The men in white coats are cunning beings. By taking distance, by turning away from the blinding clarity of the macrocosm, and by focusing their energy on the production and interpretation of inscriptions, they occupy a strategic position that controls access to the world and to discourses about the world.

If the inscription is crucial in the process of production of scientific knowledge it is because it is Janus-faced. It is a mediator that looks in opposite directions, and this is what makes it so fruitful. First, as a trace, the inscription refers to an entity whose existence is thus *(sup)posée*—that is, presumed. Something must have activated the stylus of the spectrometer that draws a peak on the squared sheet, something must have moved it to trace this perfectly determinate curve, and this something must exist. Certainly, the hand that guides the stylus is invisible to the researcher's eyes, but the signature is there, which proves the existence of the signatory. It is a signature that he must first authenticate, that is to say, compare with other known signatures, and that he will then have to reproduce in order to assure himself that it really is always the same thing that is signing. The trace registered a thousand times on the photographic plate of different bubble chambers points toward the particle one is looking for, the reality of which thus ends up compelling recognition. The inscription leads to the hypothesis of the existence of an entity. To confirm it, other signatures must be discovered, obtained in different experimental circumstances, and linked to inscriptions already provided by better-known entities. There are no better models of research in action than the little-known discipline of epigraphy. But instead of investigating inscriptions on monuments or tablets haphazardly brought to light by excavations, this epigraphy would organize its own excavations by relentlessly devising new experimental activities (*manipulations*) to produce new inscriptions and to enrich its interpretations.

The realism of the operations carried out by the laboratory is lodged in the hunt organized by researchers to identify the signatories of the traces produced. The inscriptions produced by the instruments are not any, arbitrary inscriptions. They have the weight of inescapable constraints: A well-equipped laboratory (that is to say, one that is as well equipped as its competitors) cannot make an electron, a fragment of DNA, a socio-professional category, or a W boson, say, or rather write just anything whatsoever. In their laboratories, researchers make their experimental apparatuses work realistically as if they were dealing with nature, an artificial nature certainly, but a very real nature that imposes its rules of organization. And this nature writes. To replicate an experiment is to manage to stabilize the inscriptions produced by the instruments by replicating the laboratory and following a formalized protocol. When this result is obtained, then a fact is on the way to being produced. The inscription is completely determined and this is why the scientist holding the bird in his hand well and truly takes the one in the bush (*en tenant l'ombre tient bel et bien la proie*).

The inscription gives access to entities of which it is the signature. But above all, when it is obtained at the end of the chain, after the intervention of several instruments assembled in series (the telescope, the CCD, the projection on the monitor, the algorithm for adjusting contrasts, etc.), the inscription is an encrypted message. It does not say "I am the signature, and thereby proof of existence, of the dwarf star AZ12K2003." Galileo, taking up the old metaphor of the book of nature, clearly saw the paradox: Nature is a big book, but it is written in a geometrical language. In other words, what the researcher has before his eyes are perfectly objective traces and inscriptions whose meaning must be penetrated and that must be interpreted. The inscription is both determined and enigmatic. All is not written, but, since the message is written, not everything can be said.

The inscription owes its second property to this paradoxical nature. It is because it says nothing explicit that the inscription induces speech, talk, and the statement of propositions. At the minimum: "I see a peak, there, which stands out against what seems to me to be background noise." Or "This diagram shows that the number of days lost through strikes has diminished regularly for ten years." Or, more boldly, "The gene is a deletion, which is as plain as the nose on your face." Or "What a beautiful example of twin galaxies being formed!" The inscription is infralinguistic. It is an inducement to talk. It encourages, solicits, and prepares the articulation of propositions; it is a sort of antechamber to their organization. This is why the inscription is taken up in discourses and accounts that both assign a meaning to it and rely upon it at the same time. It is through the medium of articulation that the signature is related, referred to an entity to which a name is given, an identity is assigned, and forms of action are imputed. We talk about the electron and its properties, the gene and its functions, and the working class and its alienated consciousness. The world is put into words. But the words would remain unintelligible and inexplicable if we tried to pass directly from observations to their theoretical interpretations; if we forgot the sequences of inscriptions, their multiple combinations, and the series of articulations that take them up in successive texts. There is no world on one side and statements about the world on the other, but a thick and extensive layer of interwoven traces and statements linked and connected up to each other. We call this chain of equivalences that is laboriously produced in the laboratory *translation 2*.

It is through the reasoned organization of the proliferation of inscriptions that researchers get into position to articulate propositions about the world, to reveal entities that are both real and unforeseeable. But how is this work of translation, which makes silent entities speak and write, carried out?

The attentive observation of laboratories at work holds some new surprises. It induces mistrust of simplistic answers. Researchers and technicians certainly play a crucial role in organizing these translations through the organization of practical operations designed to produce inscriptions, test their soundness, and define their meaning so as to arrive at stabilized statements: "The structure of DNA is a double helix." But they are only one of the many constitutive elements of what we propose to call the *research collective*. This is the real author of the statements and propositions with which scientific knowledge is currently identified. It is the research collective that brings together and coordinates the set of abilities that are necessary for the production of inscriptions and their interpretation. This is what must interest us if we want to understand the origin of the strength and effectiveness of secluded research.

The microcosm fabricated by and in the laboratory is a very real world, and the statements that are articulated by the research collective refer to this very real world. But words constitute only the visible part of scientific knowledge. If they stick to things they refer to, it is because they stem from a great many other things which are less visible, but just as present. The main element in the work of research actually consists in devising and perfecting instruments for fabricating inscriptions, then in stabilizing and interpreting them. What is essential in these successive adjustments is what is not said, and this is true for research as it is for any activity involving tacit know-how, dexterity, do-it-yourself, and adjustments, in short, for activities that call upon embodied skills which are difficult to transmit and are often learned on the job and by example. Even if some people might find the metaphor excessive, research is more like cooking than a highly abstract and disembodied activity. This is all the more true—and here is the paradox no doubt—when research is original and innovative. Then the research collective is groping in the dark, stumbling over the clarification of rules or procedures.

Alberto Cambrosio and Peter Keating, for example, have shown that to succeed in stabilizing and controlling the simple production of monoclonal anti-bodies, and then to be able to set out clearly the procedures to be followed, required a great deal of time and movement of technicians and researchers shuttling back and forth between different laboratories.[6] Obviously, in this process the control and mastery of human subjects and their actions are vital. Managing to get human subjects to accomplish the correct action, to carry out the collection, measurement, or experiment in ways that match, calls for a domestication of the body that can only rarely be complete. Nineteenth-century astronomers invented the expression "personal equation" to designate the irreducibly singular observations of a particular researcher and to emphasize not only the need to take this into account but also to contain it within the strictest possible limits. Providing a clear account of the operations and the events they enable one to produce passes therefore through the standardization of instruments, but it also involves the training of researchers and technicians. The most theoretical science always sinks its roots in the material and bodily practices that formed it and without which it could not be put to work and enriched. The resolution of a system of equations into partial derivatives or the conduct of logical reasoning call for as much dexterity, as much know-how learned on the job, as the perfecting of monoclonal anti-bodies. Scientific knowledge is 10 percent explicit and codified and 90 percent tacit and embodied—embodied in instruments (Bachelard: "Instruments are only materialized theories. They produce phenomena that bear the stamp of theory on every side"[7]), in disciplined bodies, in purified substances, in reagents, in laboratory animals (like *Drosophila*, or a transgenic mouse), or even in reference materials.[8]

If by research collective we agree to designate the set of elements that participate in the fabrication of knowledge, then it obviously includes human beings, researchers and technicians, who, through often heated debates and discussions, set up practical activities (*manips*) and interpret their results. But equally, and this can never be emphasized enough, it also includes all the non-humans (instruments, and so forth) of which we have just given an indicative list. This collective can be viewed from two different angles: first as a community of colleagues, then as a system of distributed intelligence.

It can be seen as a community of colleagues, or rather, of *dear* colleagues, since this is how civility requires researchers call those with whom they are in permanent competition. Dear colleagues are actually the only ones who can understand what takes place in the laboratory and give meaning to what one of them does and says. Nor do they fail to take advantage of this situation. They do not rest from looking for, and finding, the little hitch, the little bug that jeopardizes an experiment or an argument, and which sends the researcher who is challenged back to his old studies. Furthermore, this constant threat produces its effects even when it is not carried out. If your best friends' sole obsession is to demonstrate that you are mistaken, then you are forced to be prudent, to show an extreme reserve with respect to your own assertions, to advance only with measured steps, and, in expounding your interpretations, only to expose yourself with great modesty and only after having carried out every imaginable test. This constant criticism is brought to bear on inscriptions and the modalities of their production (are they really stable and reproducible?), but it is also trained on their interpretation, that is to say, on their articulation and integration in forms of reasoning and argument. The soundness of the results is due to the existence of this critical discussion.

When criticism comes to an end—for it may be that it is defused and fails to find anything wrong—agreement is established within the collective. It is tempting to say that the objectivity of knowledge is born from a simple inter-subjective agreement. But this would be an error. The objections raised are not solely the fruit of the cognitive activity of researchers and technicians. Bachelard devised a striking expression to denounce this error. If objectivity exists, he says, it is not that of the subject-object dichotomy that is so basic in grammar. And he adds: "Thus we can say that the objectivity of the neutron is, first, a response to objections."[9] And these objections, articulated by one or a number of researchers who criticize their colleagues, is anchored in inscriptions that the neutron has been induced to trace by interposed instruments. Bachelard is right: It is because it is the neutron itself that objects (or, more exactly, the neutron linked to instruments and researchers who serve as its spokesperson) that it acquires the status of objectivity.

So the research collective cannot be identified with a simple community of researchers. It is equally a system of distributed intelligence: What human beings can say and write, what they can assert and object to, cannot be dissociated from the obscure work of the instruments and disciplined bodies that cooperate and participate in their own right in the elaboration of knowledge. To understand this crucial point, let's make a detour that at first sight might seem incongruous. Let's embark on a battleship.

Edwin Hutchins has made a detailed study of the mechanisms that enable the crew of a U.S. Navy helicopter carrier to determine the position of the ship at any moment and as a matter of urgency.[10] This operation, however simple in comparison with the infinitely more complex tasks to be carried out in a laboratory, mobilizes an extended team and involves the cooperation of a number of human beings and diverse instruments. Several individuals take part in calculating the position: the sailor who takes the fix, but also the sailor who transmits the information, the one who records it in the log, the one who plots the fix on the chart, and finally the quartermaster who coordinates the whole operation. But different artifacts are just as much involved: the alidade, the telephone for transmitting messages, the chart, the protractor for measuring angles, the graduated ruler for measuring distances, and tables for carrying out conversions of units. Persons and artifacts are seen as elements of a collective, each element participating in the task of computation. All kinds of instruments play an essential role in this calculating collective. Obviously, they are tools without which the calculation could not be carried out. But above all, beyond their status as simple prostheses relieving overloaded human brains, or brains that may easily become overloaded, each in their own way actively contributes to the collective task. A chart, for example, provides already coded qualitative information, but above all it enables a series of measurements to be made that are not programmed in advance and are useful for confronting an unexpected navigational problem. The chart opens up spaces of actions and operations that would not have been possible with other artifacts and, above all, human beings left to themselves could not even have envisaged. Even if he has a chart in his head, the best navigator cannot carry out those measurements on it that are possible only with a compass, protractor, and graduated ruler! In order to show the radical difference between being and not being equipped with a paper chart, Hutchins compares this instrumented way of plotting one's position with the traditional technique of

Micronesian navigators. The latter, unlike the Yankee sailor, who thinks the boat moves in relation to fixed reference points (which have the property of being integrated in the chart and so of being legible on it), think that it is the reference points (in this case the stars) that move in relation to the boat. The chart makes it possible to get things done, to follow certain procedures of calculation and positioning that are not possible when it is not there. In this sense we can say that, like the quartermaster, the compass rose, or compass, it participates actively in the collective computation. The chart is not a simple tool because it supplies implicit, hidden information that an appropriate use brings to light. In addition, and this is easily understood, it establishes a relation—again, a modality of action—between individuals who, without it, could not take part in the collective task. The chart passes between individuals who live together on the ship and provides them with a means of communication; it gives rise to activities that it brings together and makes compatible and complementary. It establishes relations with individuals not engaged in the present action, like those who have worked out the charts, or who use them on other ships or on land to follow ships on an assignment.

The helicopter carrier is a good metaphor for the research collective. In such a collective we find a population of human beings cooperating with instruments, materials, and texts. And in these research collectives, the non-humans are as active as the charts, protractors, or rulers on board the helicopter carrier. The research collective includes all the laboratories that participate in the discussion of the results, but not only these laboratories. Through the intermediary of the instruments, materials, and substances that it uses, but also through the mediation of the researchers, technicians, and managers that it calls upon, it is further linked indirectly with a multitude of other collectives (like the manufacturers of instruments or the employers of researchers trained in the laboratories) which are not necessarily specialized in research activities. Reference to the notion of distributed intelligence enables one to distribute the skills usually attributed to researchers across a multiplicity of other actors, and non-human actors in particular. But it runs the risk of a possible misinterpretation induced by the word 'intelligence'. The reader will have understood that it is not only intellectual, and even less cerebral capabilities that are distributed, but also and above all embodied forms of know-how, knacks, knowledge crystallized in various materials, and craft skills.

Let us summarize. Secluded research carries out a first translation that reduces the macrocosm and transports it into the microcosm of the laboratory. Instead of being at grips with abundant, complex, and heterogeneous phenomena, researchers work on simpler objects that they can manipulate at leisure, surrounded by their instruments and their libraries. This translation leads to a change of scale that is also a reversal of the relations of opposing forces. The world becomes manipulable, and this becomes easier with the formation of a powerful, integrated, and equipped research collective. The abilities distributed in the collective work to fabricate inscriptions and to decipher them so as to get the entities that sign them to speak. Through *objections* thus expressed, the entities end up acquiring an *objective* existence. We call this process of articulation and objectification, *translation 2*. But however real this existence may be, it is nevertheless local, even hyper-local, since the facts are under house arrest in the collective and its laboratories. 1955: At Cambridge and in California, people are convinced that the statement "the structure of DNA is a double helix" refers to an objective reality, but what about elsewhere? The first two translations have been completed successfully, but what about the return to the big world? What about *translation 3*?

Return to the Big World: *Translation 3*
The laboratory is at the center of a research collective that organizes experimentations. The latter, through successive translations and reductions, replace the big world, the macrocosm, with the small world of the laboratory microcosm. It is here that the research collective gets to work. It organizes experimental work (*manips*), fabricates inscriptions, and translates them into propositions. While doing this, it explores worlds made up of previously unknown entities. This investigation by trial and error, by more or less skillful adjustments and tinkering, develops in the midst of uncertainties which are gradually removed to bring new possible worlds into being inhabited by new entities, produced and domesticated in the laboratory; worlds infinitely richer than the known worlds, and worlds that can reveal futures and fields of action that are infinitely more complex and diversified than those hitherto accessible. But how can these possible worlds be brought about, how can we get them out of the laboratory? This is the question that pierces secluded research. Does not secluded research risk losing everything in returning to the big world? Is the return possible, is it even conceivable?

Let us follow the ethnologist Sophie Houdart to a secluded laboratory in the Tokyo suburbs.[11] *Drosophila*—the fruit fly so common that one no longer pays any attention to it, and on (and with) which Yamamoto, a brilliant Japanese researcher, is working, with the secret hope of establishing correlations between the existence of genetic modifications and homosexual behavior—is obviously a natural being. At first sight it does not seem to be that different from its cousins living wild in the Hawaii jungle, or from those living in our Normandy farms. Moreover, in Yamamoto's eyes at least, this is what makes it so valuable. It is *a translated fly*. Its ancestors were taken, and continue to be taken, from the big world, the world in which we live, to be domesticated, worked on, and reconfigured in the laboratory, surrounded by researchers who observe it after transferring new genes to it and constraining it to reproduce and transmit the gene to its descendants. Because this genetically engineered and domesticated *Drosophila* is not completely different from the one that lives in freedom, a link can be established between what is observed in the small world of the laboratory and that which is common in the big world. But it is because it is sufficiently different from it that we can reduce the wild fly's complexity, vary certain parameters, and control some causal chains.

When it sets up the first operation of translation, science encounters the problem of the selection and constitution of the "work objects," which are distinct from the natural objects of the "colored and varied" world; they are simpler, but not completely different from these objects. These are the materials for which concepts are formed and to which they are applied. Since Morgan, the rat or white mouse, like *Drosophila*, informs and makes possible the knowledge of physiology and genetics. These objects are chosen individually according to what is to be investigated and proved, but also according to the advantages that they offer. A history should be written of these companions of our laboratories, like the history we have of the domestication of the animal species that now share our everyday life. We should have a history, like that of *Drosophila* written by Köhler, of the mice, rats, and chimpanzees, but also of the purified chemical substances, or of the reagents that indicate a solution's change of acidity.[12] Research collectives would not be able to work properly without them. But, so that the microcosm does not sever all links with the macrocosm, work objects must remain comparable, in at least some respects, with objects of the big world.

This is the reason for the anguish of scientists when they realize that they are working on species, quite clearly not just animals, which, by being domesticated, deviate too far from those in the wild state. With surgical precision, Sophie Houdart describes this feeling, which overcomes Yamamoto when he decides to establish a second laboratory, no longer situated in the urban and civilized suburbs of Tokyo, but on the edge of the Hawaiian jungle where *Drosophilae* fly free, uncontaminated by civilization and even less by those strange researchers who subject them to all sorts of tests and tortures with the single aim—and what a strange one!—of making them homosexual. The two types of fly are each as pure as the other, but each is a monster of impurity, however, when seen from the point of view of the other. The Hawaiian fly is *purely* wild and so it is impossible to study it in Yamamoto's Japanese laboratory. It brings with it impurities that decades of research work have purged from *Drosophilae* that have been worked on in laboratories throughout the world and transformed into highly standardized, reproducible, and comparable materials. But, by being purified, laboratory *Drosophilae*, in comparison with those in Hawaii, have become *impure*. The problem now is establishing a connection between the two *Drosophilae*. They are undoubtedly still comparable. But will the supposed homosexuality observed when the flies are forced to share the same room in a syringe transformed into a torture chamber prove resistant to the salty air of Hawaii, where the flies fly unfettered? To know this would require a comparison of the two laboratories: the secluded laboratory of Tokyo and the laboratory in the open air of the Pacific. Without this recalibration, nothing truly general could be said about the correlation between genetic mutations and sexual behavior. Does what holds in the syringe also hold in the big world? This is the question of the universalization of knowledge produced by the research collective.

The question confronted by Yamamoto is encountered by every research collective in the world. Once the world has been translated into the laboratory, once the research collective has carried out its experimentations, it remains to organize the return to the big world in order to describe it better, to understand it better, or even to act on it better.

Here again, the notion of translation is useful. The return to the macrocosm raises, first, the problem of the alliances the laboratory has been able to form around its research subjects. In order to mobilize the resources and support without which it would quickly disappear, the research collective must interest other actors in its enterprise. It's not important who they are so long as they have influence or money! *Interessement*, conceived as the set of actions intended to produce interest and get the adhesion of influential actors, presents modalities that vary with different times, research projects, or even disciplines. Consider the three regimes of curiosity, utility, and exactness, which formed the successive stages of an evolution ending with secluded research. In each of these regimes, the privileged targets of the actions taken to attract interest are transformed, as are the arguments that serve to convince those who contribute their support. We could no doubt write a wide-ranging history in which we would see a succession of different social configurations. For example, we would note that the personal support of monarchical power was crucial in the regime of curiosity, the king guaranteeing the existence and legitimacy of the Academy of Sciences and, thereby, the credibility of the experiments it sponsors or organizes. We would also find that science can develop only by interesting an aristocracy that forms its first and principal public. Everything changes in the following regime, that of utility. To get the backing of power one now has to involve it in programs that mobilize engineers and have a tangible social effectiveness. Industry, which devises and produces the instruments that are indispensable for achieving the objective of replicable experiments, enters the picture, and research would make no headway without its active participation. We could no doubt go further, although the materials for writing such a history are still insufficient. Some economists have recently advanced the notion of a national system of innovation to describe the set of bonds of reciprocity and interdependence that are progressively woven in the course of history between laboratories and their different partners or patrons. This would lead us to describe, as an example, the postwar French system by the central role of the big public research bodies (CNRS,[13] CEA,[14] INRA,[15] or INSERM[16]), by a system of higher education with little involvement in research, by major State programs caring little about the market, and by industrialists little inclined to manage commercial and technological risks at the same time. The important thing to get across is that the way researchers interest the society in which they live and work is strictly correlated with both the social configuration of the moment and, at the same time, the type of scientific practice that they pursue.

These contacts that are set up over the long term are connected to other, more circumstantial *interessements*. But in all these cases, *interessement* conforms to one and the same logic that, in military language, is that of the *obligatory passage point*. In order to organize the alliances it needs, the research collective must show that it is indispensable to those whose support it seeks: "To achieve the objectives you have determined, to defend your interests, to consolidate your identity, to make your size felt, come quickly to our laboratories, join our projects!" Of course, there is often a hint of mendacious advertising in these exhortations. And of course the researchers do not always have what they are selling in stock. But it would be wrong to reduce *interessement* to worthless rhetoric. If there is rhetoric, and it is clear that there is, it cannot be based forever on just promises. Sooner or later you have to show what you can do. This presupposes that there is a relationship between the objects studied in the laboratory and what those interested in its activity expect. This delicate adjustment usually involves mutual realignments, with the demands or expectations of the partners often being formulated in a convincing way by the researchers themselves, who modify their own research strategy in order to get attention.

Successful *interessement* is decided, first, in the movement we have called *translation 1*. Take the case of the abundant research that was initiated in France around the electric vehicle in the 1960s and the 1970s.[17] It all started with the initiative taken by a handful of electrochemists in launching a research program on fuel cells. According to them, their special field, which is at the meeting point of well established disciplines like physics and chemistry, sorely lacks recognition in academic institutions as well as in enterprises. That is why, they add, fuel cells are laboratory curiosities. However, the principle of their functioning has been known for a long time: They create a reaction of two chemical substances that, by combining (like hydrogen and oxygen to form water), release electrons that only have to be collected to generate an electric current. The fact remains that at the end of the 1950s very little was known about the fundamental mechanisms that, thanks to the presence of substances called *catalysts*, allow the generation of electricity. The process is mastered with very expensive catalysts, like platinum, and fuels (like hydrogen) which are not always easy to store and distribute. The American space programs demonstrated the reliability of these fuel cells in a spectacular manner. This technology, which remained confidential, was in the spotlight. The electrochemists sensed the good opportunity. Gaullist France, thanks to the young DGRST,[18] had just launched a research program on different technologies for energy conversion with the aim of ensuring national independence. Every imaginable path is reviewed, from magnetohydrodynamics to tidal energy. The electrochemists go to great lengths to ensure that fuel cells are not forgotten, and that they figure prominently in the picture of the technologies to be explored.

One of the finest chains of translation ever imagined by researchers is then set up. According to the electrochemists, nothing less than national grandeur is at stake. Not only can fuel cells produce electricity when transformed in power plants; they can also be used as a source of energy for electric vehicles. An environmental objective (electric motors avoid both chemical and noise pollution) is thus linked with the objective of national independence (France could dispense with petrol imports). The electrochemists' discourse finds an echo in political spheres. Furthermore, their request—"Support our laboratories and research on fuel cells"—attracts the attention of a multiplicity of groups and actors who see a number of advantages in this innovation and so decide to support it. Among these are influential political decision makers, concerned with the national stakes, but also industrialists who look longingly at new markets that could open up. There are, of course, some fierce adversaries, like gasoline producers, the automobile industry, or "French car lovers" and fans of motorsports. However, it is the relation of forces that counts in this kind of affair, the number of divisions each camp can put into the field. Those in favor of the electric vehicle mobilize more troops, and they are more powerful and better organized than those who resist. That is why in the 1960s the electrochemists and their allies will lay down the law. Their laboratories and research programs are transformed into obligatory passage points. Those who were ignored are now in the limelight. But they know full well that *interessement*, the translation, cannot be confined to the statement of these equivalences. Existing on paper, they must also exist in reality! With the fuel cell they have the support of the Gaullist state, the CNRS, which is obliged to follow its lead, as well as the industrialists in search of markets. But the whole construction makes sense only if they really have the cell! And to have it, they must control this little reaction that, taking place on a suitably devised electrode, will allow a handful of electrons to be released and then lovingly collected. For example, the fundamental mechanisms governing the interactions between a hydrogen atom and a platinum surface have to be understood. The translation is breathtaking. At one extreme is the nation's destiny, and at the other a series of experiments aiming to track electrons in a monotubular electrode. This apparatus represents, that is to say, reduces, realistically, not only the fuel cell, but also the policy of national independence, the nascent desire to preserve the environment, and commercial strategies of industrial groups eager to conquer new outlets. This improbable being, the monotubular electrode, constitutes a socio-technical model in which the world to which the electrochemists and their allies want to give birth is inscribed, not in the form of traces on a sheet of paper in this case, but in the materials that compose it and the form given to them. They strive to replace a France dependent on petroleum and poisoned by gasoline-powered engines, because deprived of high performance fuel cells, with an independent France without pollution, equipped with power plants, and crossed by automobiles operating with fuel cells. In their laboratories, the electrochemists are not content with controlling technical and scientific apparatuses, they are constructing at the same time the society that can accommodate them; they are working on the reconfiguration of the existing world.[19] Their strength is in having understood the entangled relationship between technical artifacts and forms of social organization, in having seen that it is impossible in truth to distinguish them. A visiting layperson who discovers the monotubular electrode, thinks he has

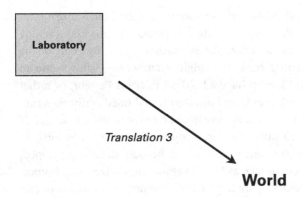

Figure 2.2
Translation 3 (transportation of laboratory results into the big world).

a purely technical apparatus before his eyes, whereas he is in front of a small-scale model of the collective, a maquette that enables action on a whole world in order to transform it. If he agreed to slip into the innermost recesses of the electrode, our observer would see lined up all those actors who are waiting to seize hold of the cell that has been shaped in order to interest them. (See figure 2.2.)

The possibility of return to the world is decided in the course of *translation 1*, in the reduction and transportation. But it is in *translation 3* that the alliances sealed by *translation 1* are revealed, and their solidity and viability tested. Will the monotubular electrode keep its promises? Is the small-scale formulation that it proposes for problems that supposedly trouble the whole of France a realistic reduction? The answer will depend on the researchers' ability to maintain the chain of translation, to hold together electrons, catalyst, public authorities, and concerned enterprises, and to resist all attempts to break it, whether they come from the electro-catalysis front, or from the political or industrial fronts. There are many dangers threatening our scientific adventurers on the return journey. They must once again change scale, complicate their models, and introduce new variables. How can this transition from the microcosm to the macrocosm (what we call *translation 3*) be accomplished without losing what was gained in the laboratory?

The answer is in a an ugly but evocative word: 'laboratorization'. For the world to behave as in the research laboratory, we don't have to beat about the bush, we simply have to transform the world so that at every strategic point a "replica" of the laboratory, the site were we can control the phenomena studied, is placed.

Let us follow Bruno Latour and his account of one the many return journeys of the great Louis Pasteur.[20] If Pasteur is great, it is precisely because he knew how to think big and completely mastered the third part of the translation, which, starting from the original laboratory, allows one to have access to the world and to have a hold on it. Louis Pasteur, or rather Roux, the doctor who took over from him, thinks that the diphtheria serum he has just developed in his laboratory should overcome the epidemic. Of course, the doctors, who until then were opposed, for good reasons, to Pasteur's work and the different vaccines that he had already developed, still have to be convinced. In fact, as Bruno Latour emphasizes with humor, "a vaccine deprives a doctor of some patients. And however much a doctor may be a disinterested type who wants to look after humanity, the fewer patients there are, the fewer doctors there are." But the research on the diphtheria serum is completely different. It translates the doctors' interests and expectations in a positive way instead of attacking them head on. The administration of a serum to a patient presupposes a prior diagnosis, and this is unquestionably a medical act that is one of the doctor's exclusive prerogatives. Here, then, is a discovery which adds to the doctor's skills and field of action, and which will consequently enrich him financially! To be able to administer the serum the doctor must agree to introduce some changes into his office, so as to transform it into an annex of the Institut Pasteur. He must educate himself, and he therefore must train himself in the methods and know-how of bacteriology. Every doctor installs a lab in his office, equips himself, and learns how to use a microscope. The doctors invest, train themselves, transform their office and at the same time themselves, no doubt judging that the reconfiguration of their skills *and* their profession *and* their identity is worth a try. This reconfiguration also benefits the patients and the Institut Pasteur, which will confirm the diagnoses and sell the serums. A network of strongly interdependent interests is formed. As game theorists would say, everything changes and everyone gains. There are some recalcitrant adversaries, but they are swept aside by the tidal wave. The laboratory has spread by reconfiguring all those who want to have it ready to hand. The difference between the world before *translation 1* and the world after *translation 2* and *translation 3* is this sudden proliferation of laboratories along with the techniques and entities that they bring with them, and the interests and projects that they authorize. One of the possible worlds starts to exist on a large scale. In passing, we can see the strategic character of the regime of utility, for laboratories first have to be replicated within the research collective before they can be launched into the big world, like Christopher Columbus's caravels, departing long ago to conquer the New World. The expression "laboratorization of society" does not mean that society is reduced to one huge laboratory, but that at different spots laboratories are implanted that frame and preformat possible actions. This movement is continuous, for not only are new spaces of action opened up by the installation of new laboratories, but those already in place are replaced by new laboratories that make the earlier ones obsolete.

Laboratorization is an interminable undertaking, always starting up again. Let us leave the Institut Pasteur to its work of transformation of French society at the end of the nineteenth century, and let us consider the genetic consultation service of a big Parisian hospital. As recently as 1970 this department did not exist. The patients brought there were, for example, those affected by forms of myopathy who, rejected by all the existing services, were looking for a serious diagnosis, or at least a name for the disease. Then the human genome research center known as Généthon opened, and genes were identified. The service began to take samples from worried mothers in order to carry out, in a central laboratory, a genetic diagnosis concerning them or the embryo they were carrying. Instead of leaving it up to researchers, the service quickly equipped itself with sequencers. It now carries out the diagnoses itself and, using its own means, is embarking on the identification of genes implicated in genetic diseases that have not yet been studied. The service is taking on researchers who are capable of working on proteins and their functions, of "screening" molecules in order to stimulate the defective genes. The laboratory is now in the doctor's office; we are no longer just replicating existing laboratories, but constructing original laboratories close to the users. Molecules come from these laboratories like the one that enables the progress of Friedreich's ataxia to be checked. Here again, the life and identity of the doctors, but also of the patients, completely changes, and even more profoundly. The clinician becomes a researcher, day by day transferring the skills acquired at the bench into the doctor's office and the treatment he prescribes; the patient, instead of thinking of himself as an isolated individual, unable to act on his destiny, knows the origin of his suffering and above all becomes capable of controlling the conditions of procreation: "Thank you doctor, I am happy to know that it is a disease and to know its name; I thank you because, thanks to you, I have had a healthy baby." The patient is gratified and, as a result, so is the doctor. He views the machines, the researchers, and the technicians—the research collective that surrounds him—with satisfaction.

This proliferation, always begun again anew, of laboratories, some of which are simple copies while others are the result of important adaptations, becomes spectacular. Automobiles are microcosms straight from industrial research centers (they are not equipped with fuel cells because the monotubular electrode did not withstand the return journey, the electrons refusing to be tamed and the oil and automobile industries vigorously striking back; such hiccups occur fairly often). The factories that manufacture CDs, or prepare the vectors for gene therapy, or that reprocess highly radioactive waste, are hardly distinguishable from the laboratories that mastered the knowledge and technology that they use. In our most trivial daily activities, we all go through constructed and disseminated laboratories, through research collectives that, with consummate skill, have been able to master *translation 1*, *translation 2*, and then *translation 3*.

Translation

The sequence of these three translations and its detour through the laboratory, situated at a distance from the big world but without having severed its moorings with it, results in a partial reconfiguration of the macrocosm, which is the transition from state 1 to state 2. We will reserve the word 'Translation' (with capital T and no number) for this reconfiguration. As is shown in figure 2.3, *Translation* is the composition of *translation 1*, *translation 2*, and *translation 3*. This is the meaning of Latour's claim that science is the continuation of politics by other means. Obviously this is not to say, and he has never said, that science is reducible to just politics, that it is only an avatar of politics disguised under an assumed name. He restricted himself to observing that, when successful, the consequence, and sometimes even the project of the detour through the laboratory, is the reconfiguration of the worlds in which we decide to live. How else, other than as politics, could we describe the movement from macrocosm 1 to macrocosm 2, the exploration of possible worlds, and the choice between them? What is at stake in this movement is actually the form and composition of the collective in which we live. What better political questions are there, what better questions concerning the forms of common life, than those concerning whether or not a society with thermal vehicles and an oil industry is preferable to a society equipped with fuel cells, or a society in which every doctor, on condition of that he agrees to train and equip himself, can struggle effectively against diphtheria or genetic diseases? Another striking way of formulating the question is to say that what is at stake is whether or not we want to form a collective inhabited by fuel cells, electric cars, motorists who have accepted them without hesitation, industries that manufacture

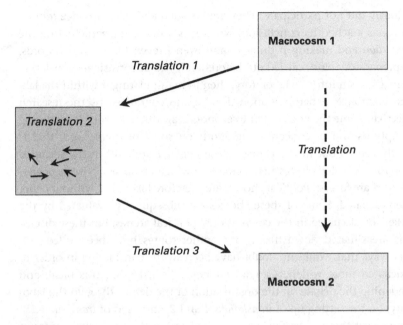

Figure 2.3
Translation is made up of three elementary translations which take the world from one state to another.

them, and ministers imposing environmental norms, or whether or not we want to form a collective inhabited by infectious agents (made visible, manipulable, and controllable), by doctors' offices able to diagnose diphtheria, by laboratories able to supply serums, or by clinical services carrying out prenatal genetic diagnoses in order to allow a woman to decide on a possible termination of pregnancy. The communities, and consequently the forms of common life, will differ according to the choices we make.

How can we fail to see that this political choice in favor of one collective or another is carried out without any real debate or consultation, that is to say, according to procedures that are not those we usually associate with political life in our democracies? The argument that amounts to saying that, in the end, the market and consequently the consumer decides, does not bear close scrutiny. As we will show in chapter 7, the market is not an abstract institution. Besides, it is inaccurate to speak of the market in general. It is better to speak of markets in the plural, and of markets which are organized and progressively structured. All the studies of innovation demonstrate that these markets are the beneficiaries of a history in which they obviously did not participate. They are installed and develop once *transla-tion 3* has reached its conclusion, when it is a matter of perpetuating the *Translation* and making it durable, and even irreversible. In other words, the preparatory work carried out beforehand, which consists in translating the macrocosm into the laboratory, then in manipulating it within the lab-oratory through experimentation devised and conducted by the research collective, frame the choices left to economic agents. The work of exploring the options and of concentrating efforts on some of them, work that is broadly carried out in the course of *translation 1* and within the secluded laboratory, develops in restricted circles in which alliances between power-ful actors are forged. Political choices are therefore taken, but without being widely debated. Some of these choices are subsequently evaluated by the markets or discussed in the conventional political arenas, but these discus-sions are situated downstream, once other routes have been ruled out, other ways that would no doubt have been worth considering in order to be debated more widely. Can we envisage reconsidering this break and broadening the debate on the organization of the detour through the labo-ratory, on the setting up of *translations 1* and *2*, and then of *translation 3*?

To answer these questions we have to abandon the false obviousness of the *Translation*. There is nothing necessary or inexorable about the passage from one state of the world to another. It is a composite operation that is not devoid of violence: violence at the moment of *translation 1*, when the specialists take their leave of laypersons; violence at the moment of *transla-tion 2*, when they shut themselves away in their research collectives; and violence at the moment of *translation 3*, when they duplicate their labora-tories. *Translation*, which appears as a progressive slide—a sort of fade in, fade out[21]—is a machine for changing the life of laypersons, but without really involving them in the conception and implementation of this change. Is this exclusion, which is no doubt one of the reasons for the pro-liferation of socio-technical controversies, inevitable? Is there such a gulf between scholarly and ordinary thought that any idea of cooperation is doomed to failure? To answer these questions, and to escape the common-places they have engendered, we will show not only that this gulf doesn't exist, but also that it is both possible and necessary to consider the exis-tence of research in the wild that is prepared to engage in cooperation with secluded research. Yes, laypersons can and must intervene in the course of scientific research, joining their voices with those of the people we call specialists.

3 There's Always Someone More Specialist

It seems that Henry IV stayed there for a night. The unpretentious building's three wings of white stone frame a large courtyard warmed by the sun. Children play under the watchful eye of their parents. You might think it was a playground, but it is not. These boys and girls are in electric wheelchairs that they handle with great skill. Nor is it an institution for handicapped children. These children suffer from a terrible neuromuscular disease: spinal muscular atrophy (SMA). They are here because their parents, all of whom live in the South of France, are taking part in a regional briefing day organized by the Association française contre les myopathies (AFM).

Since 1987, the AFM has organized a Téléthon, which mobilizes the whole of France for 36 hours. It gets people onto the streets for what seem like gratuitous ordeals but all of which aim to demonstrate solidarity with those who, owing to a small genetic muddle, no longer have the use of their muscles and, like the children playing in the sun, are condemned to a wheelchair and very quickly caught up in a difficult struggle with death. Every year the Téléthon breaks the record of the amount of money collected: about 100 million Euros. Half of this goes to help victims of the disease in their daily life, and the other half finances research whose ultimate aim is the discovery of therapies which may one day overcome these diseases that randomly wound and kill.

Such a scene would have been unimaginable at the end of the 1950s. Families hid their children out of fear of what people would say. They were thought to be suffering from an incurable defect. Thanks to the AFM everything has changed; not only has it brought the victims and their families out of the shadows; in addition, it has helped launch research on these rare and orphan diseases that fail to hold the attention of either doctors or researchers. Thanks to the Téléthon and the will of its board of directors, the AFM created a research milieu, established Généthon, and enabled the genes responsible for the main myopathies to be located and identified. The gene responsible for spinal muscular atrophy was located and identified at the beginning of the 1990s. The research team that achieved this result worked in close collaboration with families of the AFM, those that are here today with their children who we still do not know how to treat, but the origin of whose disease we do know. Since identifying the gene, the researchers have not been idle. They have worked to reconstruct the mechanisms that enable us to explain why it takes only one faulty gene for the child quickly to become unable to move, condemned to the electric wheelchair, to a tracheotomy, to painful sessions of respiration therapy and physical therapy, and sometimes even to an arthrodesis aiming to support the spinal column and allow satisfactory breathing. In this work, a notable result of which is the possibility of prenatal diagnoses that enable already affected families to avoid experiencing the misfortune again, the researchers have been constantly accompanied and supervised by the families. The latter very quickly grouped together to exchange information, share their experiences, help each other, but also to follow the work of the researchers, sharing in their successes and failures.

While the youngest children play in the courtyard, the briefings take place without break in the rooms hired by the association. This morning it is the Goussiaumes who lead the way. For a long time, Patricia and Alain, wife and husband, were organizers of the spinal muscular atrophy group within the AFM. Two of their children, twins, died some years apart, the second recently. Nevertheless, they continue to be involved with the group, a bit in the background, but always there to welcome new distraught parents. The Goussiaumes know everything about this terrible disease. It was they, with some others, who formed the group in 1986. They brought it to the attention of the medical world, made contact with researchers, and accumulated tons of information on the different forms of the disease, its development, and the different treatments to be followed and their relative effectiveness. They took an active part in the collection of blood samples to enable researchers at the Hôpital Necker—Enfants malades in Paris to locate the gene. They conducted inquiries in the families. Right at the start, when very little was known about the disease, they wrote a report of more than 300 pages in which they described all the actions to take to prevent the young patient from suffering too much and to delay the progress of the disease. Then they followed the work of the researchers, visiting their laboratories, organizing briefing sessions, and reading the scientific articles published in the major international reviews. They became experts on the disease.

Alain Goussiaume opens the meeting. He begins with a history of the group and of the disease, but also of the research. It is a mixed history, combining photos of children whose bodies illustrate the attacks of the disease, images of the Téléthon, reminders of the actions undertaken by the association to mobilize researchers and doctors, of the genes glimpsed, located, and then identified, of scientific colloquia, and copies of articles. It is impossible to separate out the different components in order to recount one purely scientific history, another purely medical history, and another history of the struggle of the families and the sufferings of the children and their parents. Each thread is interlaced with the others.

Goussiaume now tackles the second part of his presentation. After showing the seamless nature of the fabric and that the history of the discovery of the gene and of its many modifications was mixed with that of the patients and their efforts, and that of the association and of laboratories, he zooms in. He follows one of the threads of the macramé that enables him to show that this history has a direction, and that, as everyone hopes and wishes, it will result in the medications and therapies they are waiting for. "This history," he declares, "is a book with five chapters. It is a book with a complicated text." The presentation takes a more technical turn. The public, which includes not only the families, and some patients in their wheelchairs, tired by the games in the sun, but also researchers and doctors who are specialists in the disease, will at no time let themselves be distracted by the cries of the children playing in the courtyard. Alain makes an obscure history of genes and proteins crystal clear. Finished science, cooled down, the science taught in colleges and universities, is often daunting and without interest; science in the making, hot, hesitant, uncertain, moving from one problem to another, sometimes discovering fruitful tracks, at other times losing its way on shortcuts, the science called research, is fascinating, especially when it speaks of those to whom it is addressed. Goussiaume knows that he must be careful. He faces a mixed and heterogeneous audience, parents who know nothing about biology, but also the best scientists of the time. Hence he chooses his words carefully: "If I say something stupid, don't hesitate to correct me!" Goussiaume does not forget for a moment that he is not a specialist; he is qualified in architectural design, a domain far from genetics! But he is the father of Rémi and Marc, the twins with whom he has accompanied the disease for more than ten years. In his presentation, Goussiaume spares us nothing. We get positional cloning, microsatellites, promoters, strategies for modeling the disease, and hopes for medications. Then Goussiaume starts on a course in molecular biology, starting from the basics but "going to the essential," all illustrated with magnificent computer diagrams: "We think that there around 100,000 genes," he says.

"This has just changed! Since last week it is thought there are 160,000," exclaims a scientist in the room.

"No! There would only be 35,000," says another.

"Oh dear!" groans the audience, stunned by all these figures.

Goussiaume then moves on to the different types of spinal muscular atrophy, to the evolution of the classifications, the clinical symptoms, and to correlations between forms of SMA and genetic characteristics. Presenting the most recent results, he concludes: "There are four different proteins and therefore four different forms."

"That's interesting, I didn't know that!" a third scientist exclaims.

Passing to the description of the role of the protein in the cell, Goussiaume clarifies its link to the second stage of the development of the RNA messenger, that of maturation. Since he thinks that with this proposition he has arrived at the very latest results from the research front, he asks: "Do we know any more about this?"

"The model has changed a bit, but we are not sure about it," a researcher replies.

"Even so, Fujimoto has described another function," Goussiaume emphasizes.

"Could I say something?" a researcher asks. "This is completely false. No one has been able to replicate it."

"Nevertheless, an article appeared in a very good journal!"

"Yes, but that's the kind of thing that happens."

Then a discussion gets going between scientists working on the muscle and those working on the motor neuron, on the subject of the interaction between the two. We are at the heart of ongoing controversies, hesitations, and debates between specialists. And the confrontation takes place amidst parents and their children affected by the disease, under the interested eye of Alain Goussiaume, who thinks he will be able to form an opinion. He does not allow any element of information to pass: "I am not sure I have understood. Could you explain it again?"

This scene, astonishing as it may be, is not extraordinary. In the AFM, many sufferers or their parents become experts of their disease, able to follow the work of researchers, and also able to cooperate with them on some points, having played a central role anyway in the primitive accumulation of organized and formalized knowledge about the diseases. It is not only in the AFM that this takes place. After following the struggle of associations of people with AIDS for a number of years, Steven Epstein has made the same observations: some patients become experts among the experts; others do not hesitate to propose new forms of clinical experimentation.[1] In a later chapter we will see what is original and innovative in the participation of laypersons in the development of knowledge that concerns them. However, before coming to that, we must devote some time to understanding how laypersons, non-specialists, succeed in entering into dialogue with the best experts, sometimes even suggesting ideas and making proposals to them. Of course not all patient organizations are based on the AFM model—far from it. But the AFM's history shows that this type of adventure is possible. It causes us to wonder how we can account for this irruption into the world of research by those who previously were carefully kept at arm's length. Do they really contribute something? Or are they merely tolerated for diplomatic reasons?

It is easy to imagine the fears aroused by such an intrusion. If the experiment is continued and broadened, will not science risk being harmed by the hordes of barbarians who come into the laboratories to bother those who need peace and quiet? Dialogues and exchanges that provide researchers with a favorable environment are fine, but only so long as they change nothing in the regime of the production of science and in the modes of organization of research! Alain Goussiaume, like all those who become experts among experts, is welcome. However, he can only be, and must only be, an exception, someone transformed by circumstances into a good student that specialists welcome with pleasure. Beyond this slight enlargement of the circle of interested actors, nothing else must change: the genes and proteins are there, and they impose their distinctive reality on researchers without waiting for patients to be interested in them. These meetings, the efforts parents agree to in order to penetrate the world of research, oils the mechanisms, but it does not change them. Science is a matter of laboratories, nothing more and nothing less.

What if things were not so simple? Certainly no one imagines that genes will change, as if by the wave of a magic wand, when patients come on the scene. More simply, the question is one of being prepared to consider the possibility that the way of formulating problems, constituting the research collective, and then disseminating and implementing the results may result in a different form of organization and integration of research in the social fabric. We ask the reader who thinks he can see in this suggestion the deadly poison of irrationalism, and who thinks he hears the discourse of postmodern relativism, which in attacking science attacks the foundation of our civilization, to grant us a little of his time and attention. We ask him to be willing to consider that we are not assembling a war machine against science, but rather that, benefiting from what are now numerous experiences of cooperation between specialists and non-specialists, our sole ambition is to show that there are other ways of doing research than those that have been laid down over time.

To bring out both the necessity and possibility of these collaborations, we must forget for a moment the existence of the great division. We must be willing to acknowledge that laypersons and specialists are engaged in research activities and that at certain moments they reckon it is a good idea to combine their efforts. The preceding chapter indicates these meeting points. Exchanges and collaborations may be established at the point when problems are formulated and the professional researchers are about to enclose themselves in their laboratories. What is at stake here is the content of *translation 1*. They take place again when it is a question of organizing the research collective and managing how it works: *translation 2*. Finally they arise at the moment of return: *translation 3*. By focusing our attention on these critical episodes we will manage to elucidate the different modalities of collaboration between specialists and laypersons.

Taking Part in the Formulation of Problems

The history proposed by Christian Licoppe shows that the transition from *experientia* to *experimentum*, from common knowledge to knowledge mastered by the research collective alone, is effectuated only if some unexpected questions, some surprising, disconcerting phenomena stand out and hold the attention against the background of shared knowledge. These anomalies challenge available knowledge and convictions and give rise to the formation of new forms of knowledge. Through the surprise they provoke they constitute one of the mainsprings of the regime of curiosity. Something happens, an unheard-of event, which, once the gaze is fixed on it, imposes itself as an irritating question that poses problems. Invisible things enter the field of perception and upset habits of thought as much as practices. In his study of rationality, Robin Horton notes that at the origin of Western science is an unhealthy attraction for the monstrous, for everything that contrasts with habit, everything that puts established regularities in danger.[2] Explaining the unexpected, revealing the causal chains that enable us to account for the new, is one of the obsessions on which scientific knowledge is constructed. The profession of the researcher is to track down monsters and bring them out of the dark in which they are hidden.

Speaking of monsters or of problems is roughly the same thing. The monster we display, and which intrigues us, is like an obstacle on the road that we usually take, blocking our progress. The starting point of any research undertaking is the fabrication of true problems or the identification of phenomena that pose problems. With no problem to resolve there is no incentive to produce new knowledge. We can situate a first, active contribution, a first point of entry of laypersons into the process of the production and dissemination of scientific knowledge in the work of giving prominence to problems, identifying obstacles, and revealing strange and bizarre phenomena. Problems, in fact, are not the monopoly of experts.

Take the case of what Phil Brown calls "popular epidemiology."[3] Contrary to the traditional view that contrasts lay and professional forms of knowledge term by term, what took place in Woburn, Massachusetts, in the 1980s shows the complementarity of the two approaches. Over a period of years, the residents of this small county notice the occurrence of a large number of infantile leukemias in their families; the fact that these observations are made by several groups of people independently of each other gives them some solidity. This oddity, this monstrosity Horton would say, poses problems. It attracts the attention of the population and becomes worrying. Faced with the exceptional, with the unexpected singularity, there is naturally a search for explanations. In such a case, the investigation, whether conducted by laypersons or specialists, endeavors to reconstruct the causal chains. Edward Evans-Pritchard, in his admirable study of the Azande, has shown the tension at the heart of these investigations.[4] That the spear kills the warrior is easily explained, or at any rate, can be easily integrated in *experientia*, in common knowledge: a wounded body that loses its blood is a body that sinks into death. But that this particular spear struck this particular warrior, and not another, is a singularity that no common knowledge can account for and that the Azande nonetheless want to explain. They say that the spear kills twice, once in a contingent manner (it happened that this body was on the trajectory of that spear), and again in a necessary manner (the sharp iron caused the irreparable damage). It is the contingent death that poses the problem. Faced with this enigma, the Azande are not content with invoking some kind of fate, as a superficial Western mind would be inclined to think. They must understand why the tree from which the spear was cut was felled, and why the warrior armed with the spear crossed the path of the one who was to become his victim. Like you and us, the Azande need solid causation; they have discovered none better or more plausible than those that originate in the ill will of a sorcerer, who must be identified. Jeanne Favret-Saada says the same and adds that this search must take account of the field of forces formed by relations of power: some are more likely to be accused than others.[5] Sorcery is a sophisticated theoretical construction engendered by exacerbated causal thinking. Where the physicist and biologist give up, thinking that there is nothing to be understood, the ordinary person wants to understand, striving to reduce the occurrence of singular events to plausible explanations. The stubborn will to trace back the long and improbable series of causes and effects is not the monopoly of men in white coats; it is the commonest thing in the world. When the expert abandons the investigation, powerless, the layperson bravely continues with it.

Medea, act one, scene one: "Would that the Argo had never winged its way to the land of Colchis through the dark blue Symplegades! Would that pine trees had never been felled in the glens of Mount Pelion and furnished oars for the hand of the heroes who at Pelias' command set forth in quest of the Golden Fleece! For then my lady Medea would not have sailed to the towers of Iolcus, her heart smitten with love for Jason, or persuaded the daughters of Pelias to kill their father and hence would not now be inhabiting this land of Corinth with her husband and children."[6] Euripides' tragedy opens with this long lament of Medea's nurse. Medea is mad with jealousy because Jason has decided to marry the daughter of Creon, king of Corinth. The servant has a foreboding of the drama, Medea's murder of her own children. How can the inexplicable be explained, the most monstrous act there may be, taking the life of one's adored children with the sole aim of making the unfaithful man suffer? We think of the common explanation, which Euripides repeats throughout the tragedy: what happens is the ordinary drama of ordinary jealousy. But why has this drama taken possession of Medea and Jason? That is the question. To answer this question, the servant goes back along the chain of events, seeking the first cause of the tragedy in the forests of Thessaly: the tree in the woods from which the oars were cut that enabled Jason to undertake his fine voyage and to throw himself into this history full of frenzy and blood. Like the spear, jealousy kills twice. We need two explanations, and the servant knows this. Laypersons, we were saying, are infinitely more demanding than specialists when they come across a problem which resists them, especially when it is an existential problem. Especially when it involves illness or death that seems to strike at random. Phil Brown does not say this, but we can easily understand that sorcery develops in societies that have opted, against secluded research, for what we propose to call research in the wild. It allows the conditional past—the mode and tense invented to go back in history and grasp its bifurcations—to be replaced by the conciliatory certainty of the indicative and its simple past.

The inhabitants of the Massachusetts city of Woburn are in the same position as the Azande or the peasants of the Norman bocage. All are faced with a series of misfortunes that strike them in their flesh and blood. How can we account for the death of someone close to us? How can we explain the fact that entire herds of cows no longer produce any milk? How can we make sense of these leukemias that strike down innocent children? If all these rents in the weave of multiple human lives do not find an explanation, then they end up making the lives of those they wound absurd. The inhabitants of Woburn could have opted for witchcraft as the explanatory theory. No doubt they would have done so, if they did not have professionals ready to hand, if they did not have access to laboratories, and if they had not had the opportunity to sweat over chemistry and biology textbooks at school. Witchcraft is a pure form of popular epidemiology. And nothing is any longer pure in the Commonwealth of Massachusetts. What the inhabitants of Woburn want to develop is a hybrid, composite form. It is a compromise between pure knowledge in the wild and pure laboratory knowledge, and that is why it is of interest here.

Faced with the question "Why do our children die and not those in the next city?" the families throw themselves wholeheartedly into the exploration of the causal chains, into an outdoor epidemiology with the sole aim of establishing connections and revealing the relations of cause and effect. They very quickly come across the presence of industrial dumps and the existence of pollutants. Then nothing holds them back. Like good researchers, like good investigators, the parents hypothesize that this monstrous epidemic is due to the presence of these pollutants and their effects on their children's health. The residents talk about it and join together. They form a community which is no longer the community of peaceful citizens sharing the same territory and managing the same local institutions, but a community that has integrated into its daily life the presence of pollutants that take part in collective life by acting day by day on the health of the inhabitants. Industrial waste was external to the community, expelled from the collective, confined to dumps that ended up being invisible; here it is in the field of vision once again, a full member, for worse rather than better, of a collective that becomes aware that it has been living with it unknowingly for a long time. The group becomes more cohesive, as if sudden awareness of the presence of toxic waste had strengthened the social bond and produced solidarity between individuals who previously were weakly linked. The people start reading, asking questions, and exchanging information. They talk to officials, meet scientific experts, and endeavor to acquire knowledge about the supposed effects of toxic waste on the health of the residents.

The machine is launched. The residents make the search for causes their own cause. They highlight new facts, establish correlations, and construct a database that obviously did not exist. They grab hold of government experts to whom they pass on their information. The latter conclude (should we be surprised?) that there is nothing strange or monstrous. "Move along, there's nothing to see. All these supposed problems are collective hallucinations." And as these experts, while being experts, are nonetheless human beings, they add "We understand your feelings and we share your pain." Who understands whom? The group thinks it understands that the official experts do not understand anything. It appoints its own experts and initiates legal action. The case is re-opened. The studies are begun again. A public debate is organized. Hypotheses and methods are widely and openly discussed. Months pass and the results begin to accumulate. A register of cancers is set up; a five-year study is launched using retrospective and prospective data, research on genetic mutations caused by trichloroethylene is financed by the Massachusetts Institute of Technology. The residents' action is not confined to the organization of a rigorous investigation. The families ensure the continuity of the program, the quality of the measurements, the coordination of initiatives, and the formulation of new questions and hypotheses. In short (taking up a terminology from public works), they play the role of contractor and project manager, devising the programs and managing them. There is nothing more stubborn, more painstaking, more careful, and more rigorous than a group of non-specialists who want to know why they endure unbearable misfortunes. These qualities are profitable in strictly scientific terms: the result is the discovery of the trichloroethylene syndrome, which involves, at the same time, the immune, cardiovascular, and neurological system, a syndrome which, once identified, becomes prominent in other places.

This story, one of many, shows that it would be absurd to contrast lay knowledge and expert knowledge by resorting to terms like rationality and irrationality, objective knowledge and subjective beliefs. In this case, the opposite is true. Conservatism, pusillanimity, absence of intellectual openness, and the refusal to welcome the unexpected are all found on the side of the experts. Boldness, attention to novelty, and the spirit of innovation are the qualities found in the laypersons' camp. The confrontation initially takes place on reversed fronts. But symmetry is soon restored through the grace of the families, who do not want to humiliate the experts in the arrogant way that some experts sometimes take pleasure in humiliating non-specialists. They strive to create a common front against the misfortune, to enlarge the research collective, and to establish cooperation between equals. The residents do not want to work against the researchers, but with them. And it is because they deliberately adopt this perspective of collaboration that they succeed in drawing attention to unnoticed past events, arousing interest in their problems, and getting these problems taken sufficiently seriously for wide ranging investigations to be decided.

This collaboration has undoubtedly never gone so far as in the case of neuromuscular diseases.[7] Is not the basic requirement of scientific inquiry— namely, to display monsters in order to absorb them in regularities which transform them into ordinary events—literally applicable to those children suffering from myopathies which contort their bodies and in some cases affect their mental and intellectual faculties? Until quite recently families hid their children, being unable to bear them leaving the private sphere, the desire to hide them sometimes even pushing them to avoid their graves being too visible or easily accessible. The doctors, the great majority of whom were unable to give a name to these diseases, or even to propose palliative treatment, paralyzed before these families paralyzed by anxiety, found no other words than these: "Do not get too attached to them, for fear of pointless suffering, because your children are lost." For everyone these children were, in a way, monsters. Was it not said that they were defective, tainted, and so different from others that sometimes we hesitated to consider them full human beings? But they were monsters that were hidden, instead of being exhibited so that they could be studied. The painful but crucial decision, made by several families at the same time, was to show the monsters and make them exist as problems. Thirty years later, those suffering from myopathy take part in television programs and are interviewed by journalists interested in their lot. At the same time, they are at the heart of both clinical and biological research programs. And science has done its work, putting what was shocking in its singularity back in the perspective of causal determinations. The problem has still not been resolved, but the monsters have disappeared because we became interested in them, because they were recognized as the source of questions to which answers had to be found.

This movement to scientific problematization would have been impossible without the families. The creation of an association and campaigns were needed to create awareness in public authorities and medical institutions. But above all, as in the case of the inhabitants of Woburn, the sufferers and those close to them had to make a real effort, that is to say, take responsibility for part of what we can call the primitive accumulation of knowledge. In fact the first stage in the long process of transforming a monstrous phenomenon into an ordinary, common, expected phenomenon consists in drawing up an inventory of the monsters, comparing them, and grouping them into families according to their similarities or dissimilarities. This work of classification comes just after the monster's public showing or "monstration" which in itself only emphasizes and repeats its singularity. To have done with this and get scientific investigation underway, these singularities must be set alongside each other so that the first, previously invisible regularities leap to the eyes of the least informed. This is what the residents of Woburn did by creating a database to show the repetitive character of cases of leukemia. It is what the families of those suffering from myopathy did by throwing themselves into wide-ranging inquiries that enabled them to record cases and collect standardized information. They also use what we can call proto-instruments which enable them to establish indisputably the trajectories of the disease's progression, show that the children live to a certain age, and follow the stages of the disease. Thus, films are shot and photographic albums compiled which are not solely intended to fix the fleeting moments of a life that is passing away. Both films and photos contribute information and make rigorous, objective, and repeated observations communicable. The families, like those of Woburn, give shape to what will become their *experientia*, their common experience, which did not exist before, since each of them was living cut off from the others. On the basis of this *experientia* it will be possible to devise and conduct *experimenta*, or experimentations. Laboratory knowledge cannot thrive on sterile land. It would have quite simply been unthinkable without this first basis, this nourishing and fertile compost, without the monsters being displayed and then reduced to some initial regularities. Here again, as in the case of Woburn, the movement corresponding to problematization draws its energy from a close mixture of passion and reason. The film, worn out through being watched, and which shows the inexorable progression of the disease, brings the children back to life and revives the pain of their loss at the same time as it describes with clinical precision the picture of the symptoms and their development.

At Woburn, as in the case of myopathies, scientific research would have been impossible without the work of laypersons, quite simply for lack of an object, of a problem. Take away its problems and science disappears; supply it with problems that it had not seen and it is enriched. We have seen the role played by families in formulating and revealing new problems in which laboratory researchers gradually became interested. Let us generalize. Let us call concerned groups those groups that, alerted by unexplained phenomena which concern and affect them, decide to make problematic events visible and undertake a primitive accumulation of knowledge. We have to acknowledge that these concerned groups are becoming increasingly present on the public stage, but also increasingly loud and active, with the multiplication of debates concerning the environment, health, or food safety. Their role appears to be crucial and irreplaceable. In fact, the cycle of the production of knowledge would not get started without this popular epidemiology. There is a contraction of what, in the history described by Licoppe, was extended over a long period. The arousal of curiosity is constantly being revived. New problems continuously arise from everywhere. What some well established scientific disciplines have ended up forgetting—namely, that they were caught up in a trajectory that began with the public discussion of disturbing problems—is becoming the daily bread of research and researchers. Do we need to multiply the examples and go back over the history of reproductive technologies,[8] so well recounted by Adele Clarke, which shows that without the passionate, stubborn, and able involvement of women's associations, these technologies would simply never have undergone the development we have witnessed.[9] We fear this would tire the reader. We hope we have convinced him or her of the existence of this first possible point of entry for laypersons in the cycle of the production of science: that of problematization and the primitive accumulation of knowledge it requires. It is now time to move on to the second moment of the involvement of laypersons in secluded research.

Taking Part in the Research Collective in Order to Broaden and Organize It

The laboratory, or rather what we have called the research collective, is the second point of entry for laypersons into the process of production of scientific knowledge. In some circumstances, non-specialists and, more precisely, concerned groups, often allied with experts or researchers, enter the scientific arena itself, taking advantage of controversies underway in order to intervene in the debates and emphasize their points of view, concerns, and perspectives. They sometimes demonstrate that laypersons can find a place and make their voices heard at the heart of research, in its most technical and esoteric compartments. The recent and most striking example of this is the movements of those with AIDS, and a further illustration is provided by the involvement of some groups in the measurement of radioactive effects.

In the mid 1980s, different associations concerned with the treatment of AIDS begin to develop actions aiming to make up for what they consider to be failings of official institutions: "To rely solely on official institutions for our information is a form of group suicide."[10] Struggle against the authorities is organized. In the case of AIDS it finds a favorable terrain. In the first place because, as many observers have shown, at the end of the 1970s and the start of the 1980s, there is an upsurge of social movements focusing on private, even intimate aspects of human life. These social movements—like those of feminists and homosexuals, or anti-psychiatry—mobilize against the normalization of existence, that is to say, against the imposition of rules, categories, and interpretations developed by authorities external to the individual. The movements linked to AIDS build on the foundations of gay and lesbian movements to defend an identity that they see as being denied or stigmatized.[11] This protest goes through a frontal attack on medical power. It is easy for it to find a place in the public domain inasmuch as the associations involved bring together many doctors, intellectuals, and scientists, as well as people in the prime of life, physically active, and strongly determined to do battle.

In the United States, the associations begin by importing unauthorized medication, and soon realize that they will not be able to influence how the institutions function if they confine themselves to demonstrating in the streets of Manhattan or shouting slogans in front of the green lawns of the White House. To change the course of science and technology they have to enter the arena, sitting down at the experts' table, even if uninvited, and to do this they must become credible.

How do you become credible? First, by becoming competent, by taking part in conferences, by dissecting research protocols in order to acquire mastery of the technical vocabulary, by tracing medications back to the laboratories from which they came, and on the way by inspecting the work of the firms, administrations, and public bodies that have been involved. Gradually, some patients become specialists with recognized skills, real interlocutors for professional researchers.[12]

But being competent is not enough; to represent you must also possess legitimacy. Becoming an expert among experts is one thing; making sure that you continue to speak in the name of the base, that you really are the movement's spokesperson is another. In the case of AIDS, continuing to be representative is difficult, because the population of patients is varied, with multiple and even contradictory interests and aspirations. There are more dissimilarities than similarities between a black HIV-positive drug addict from the Bronx and a white HIV-negative homosexual from Greenwich Village.

The link between these two requirements—active participation in the research collective and maintaining the link with the social movement—will be made around the notions of representation and representativity, which are at the heart of clinical trials. How in fact can we decide on the effectiveness of a new molecule? The usual response is in the strict observance of a protocol that was established and codified long ago, which specifies the rules for the recruitment of patients or non-patients on which the molecule will be tested, and which also imposes double-blind trials. For these trials, two populations are formed, one on which the molecule is tested, and another which is given a placebo, that is to say, an inactive molecule. Furthermore, neither the patients nor the doctors who ensure and follow the treatment know whether they are dealing with the molecule being tested or the placebo. These rules have been worked out over time with the aim of ensuring the objectivity of trials. They aim to eliminate subjective bias, of both patients and doctors, which, given the importance of mental and psychological factors in the domain of health, may influence the effectiveness of the treatment. However, in order to be objective they are potentially in conflict with some ethical considerations. First, the notion of representativity hides choices that pertain to pure and simple morality. Why, for example, exclude certain categories of patient? May not lifestyles, attitudes toward treatment and the disease, and biological characteristics vary according to different groups? Why are Afro-Americans, women, or other minorities often under-represented in clinical trials? These questions are inseparably scientific, political, and moral. From an excessive desire to regard any sick body as equivalent to any other sick body, we might end up not being able to grasp differences of effectiveness and preventing some groups from benefiting from the possible chance of cure or remission offered to those selected for the trials. A related question arises when first indications seem to show that the molecules tested are effective. Should we then continue to administer placebos to patients, who are thereby denied the chance of seeing improvement in their condition? Some patients' associations give radical answers to these two questions. It is intolerable, they say, to keep minorities out of the scientific investigation, thus excluding them a second time, just as it is intolerable to continue with placebos when we know that treatment is available. Scientific objectivity, they add, does not call for letting someone die solely so that we can be sure that one of his randomly selected companions in misfortune may be cured or see his condition improve!

The associations are involved in all these and other aspects of research, developing arguments and pushing for more thorough investigations on some subjects. They also insert themselves in ongoing debates or revive old ones. Take the case of the controversy on the way to conduct tests of drugs. There are two opposed points of view among specialists: the pragmatic conception and the purist conception. For the first, trying to purify the protocol, that is to say, requiring only subjects who have not followed any other treatment prior to the trial, is unrealistic. To assess the true effectiveness of a new molecule, they claim, we must get close to reality, which is never pure but always disordered: medicine never deals with patients free from all treatment. For the second point of view, to get the "right" answers we must, on the contrary, purify the protocols as much as possible, selecting patients "not polluted" by previous treatments. This leads to the elimination of some patients, and of populations of patients in fact, which, in the eyes of the leaders of the associations, means de facto segregation. This debate, which cuts across the scientific and medical community, is exemplary; in it, we catch hold of the tension between laboratory research, which wants to work on purified objects, and research in the wild, which is faced with composite, impure, polluted realities. Another issue emerges beyond this opposition. When knowledge acquired on purified objects looks for its areas of application in the real world, and when it strives to maintain the conditions of its effectiveness, it is bound to demand that patients must themselves be "purified" if they want to be cured. The enterprise of normalization is not far off! And the associations are worried about this. Not all of them. The patients' spokespersons are divided. Some become more radical than the most radical specialists and call for the existing protocols to be strengthened, because, they say, only in this way can we really know the value of a molecule; others give their support to those defending the interest of a more pragmatic approach. The *Kriegspiel* can begin.

If we wanted to give a more complete picture, we would have to refer to the debates around biological markers, such as the CD4s, antibodies whose presence or absence enables one to follow the progress of the disease and decide whether or not a treatment is effective and whether it is preferable to begin treatment during asymptomatic seropositivity or wait for the disease to appear. The associations give their views on this subject as on many others. They compile different results, highlight contradictions, and stigmatize those conclusions they judge to be hasty. In any case, once they have entered the research collective, the patient, or rather their spokespersons, do not leave it. They are found writing articles published by academic journals; they take part in financing decisions based on scientific expertise; they contribute to the development of new regulations. All in all, what is at stake is the formation of groups that simultaneously assert their identity, fashion it in action, and demand new types of relations with the professionals, demanding, even if it means paying the price in terms of training, to be actively involved in the definition of the orientation of research so as to emphasize their points of view and conceptions. We are far from the judicial model in which laypersons hire the services of counter-experts—transformed into mercenaries—to whom they delegate the defense of their interests and sensibilities. These patients are concerned groups that, through intermediary representatives, get a foothold in the research collective, which is thereby broadened. What is at stake is the scrambling of the distinction between the object of research (the disease) and the subject of research (the patient who wants to be cured). Subject and object merge in the same person.

The involvement of laypersons in the research collective may be exercised without their physical presence. We shall consider the programs for the storage of radioactive waste. One of the options studied, and in truth the only option reckoned to be feasible until recently, is deep burial. Containers of waste are deposited, for the order of several hundred thousand years, in a gallery dug at the bottom of a shaft several hundred meters deep drilled in what are thought to be safe geological strata. With the passing years, as the project became clearer and the names of the burial sites began to circulate, reservations and then open opposition emerged. Although the experts were definite, too definite in the eyes of some, were we really sure of the absence of a risk of contamination of the biosphere? Or, to be more precise, could we commit ourselves on plausible detailed assessments and trustworthy probabilities? A painful question! When it is a matter of storing highly radioactive substances, which will remain radioactive for such a long time, how can we be sure of anything at all? How can we control all the variables and parameters when even the geology cannot be considered as a stable reality? The real world or the macrocosm, in which the parcels are to live for thousands of years, is so complex that it is difficult to model it. The multiplicity of interactions and the diversity of the phenomena are such that no research program in the world can claim to absorb all of them. The Earth has its history of which man is ignorant. Will the clay of the Champagne plateau remain inert and play its protective role? Will water seep through the reinforced concrete or the tempered steel? Will geological accidents occur? Once all the possible causes have been analyzed, once the chemical or physical micro-phenomena have been studied, it remains to organize simulations on powerful computers. Since the world cannot be "laboratorized," the scale of the world exceeding that of any imaginable laboratory, then it must be simulated without forgetting or concealing anything.

The credibility of models and simulations depends, first, on taking all the relevant variables into account. But this is not all. It also depends on the form of the equations. Different mathematical formalisms will have to be used depending upon whether the phenomena is considered to be linear or asymptotic, with threshold effects or without threshold effects. The following stage, digitization of the equations, is no easier to cross. Generally the resolution of systems of equations permitting the simulation of complex phenomena is obtained only by recourse to digital algorithms based on successive approximations. Behind each system of equation, behind each method of resolution of these equations, there is therefore a hidden model of reality, a certain simplified representation of the phenomena studied. The natural temptation of the engineer or researcher who implements these programs is to proceed sequentially. First, they pose the question of the appropriateness of the mathematical model and carry out a number of simplifications they judge to be plausible and acceptable. Then they move on to the digital model and with simplifications whose value they evaluate by assessing the distortions they introduce with regard to the mathematical model (itself simplified). Then they attack the computer coding, the choice of data, and so on and so forth. At every stage they simplify the hypotheses of the previous stage. It may not be the same researcher who effectuates the approximations at different stages, for the skills required or the volume of work necessary are beyond the capabilities of a single individual. As in the game of Chinese whispers, we sometimes find ourselves at the end of a chain with simulations that no longer have any relation to the original model, because each stage introduces divergences that combine with each other to make the final simulation completely unrealistic.

The vigilance of the researcher, or more exactly of the research collective, can be maintained only if it is constantly called to order: Isn't this new simplification in radical contradiction with the initial hypothesis? Doesn't the choice of writing linear equations, for reasons of simplicity of calculation, introduce wide margins of error? Keeping the implicit hypotheses underlying the formalisms adopted constantly in mind at every stage of the simulation is an undeniable guarantee of validity. The research collective has no reason to impose this requirement on itself spontaneously, especially when internal debates between colleagues are discouraged on the grounds of confidentiality. On the other hand, the research collective's critical spirit will remain on the alert when under the watchful eye of someone external who demands explanations for the simplifications made at each stage. And the pressure will be even stronger if laypersons, in seeking to be listened to better, resort to external specialists who stimulate exchanges. The credibility and influence of laypersons is greater if they have access to facilities which enable experiments and analyses to be replicated. This solution is particularly effective when, for historical reasons, the research collective is turned in on itself and so secluded that no internal discussion or criticism is possible. One example among many is the Chernobyl accident in 1986. It will be recalled that the French experts, relying on measurements available only to them, asserted at the time that the radioactive cloud, no doubt discouraged from entering our territory by the formidably effective Customs control, had complied and obediently agreed to bypass France. It needed the mobilization of groups outside this confidential, because restricted, collective for the results and hypotheses to be discussed. This was possible owing to the creation of a research centre (the Commission de Recherche et d'Information sur la Radioactivité, abbreviated Crii-Rad), outside and independent of the collective that had monopolized expertise on the subject.

Non-specialists can therefore take part in the research collective, in the debates firing it and in the choices that it makes. Sometimes this participation is direct, as in the case of AIDS. But, as in the case of nuclear technology, it may equally be indirect, whether through the vigilant presence of concerned groups, which fosters greater prudence and professional consciousness on the part of the researchers, or through these groups calling upon experts in order to exercise this vigilance and create a space for discussion.

Under these conditions, why should we not regard laypersons, whether or not they are allied with experts, as acting as genuine researchers in the wild when they join with these and demand, if it proves necessary, greater rigor and rationality in the management of the production and interpretation of inscriptions, which we have seen constitute the material on which laboratories work? Just like *translation 1*, *translation 2* may be enriched by the involvement of concerned groups.

Turning Back to the World

The third point of entry for laypersons is situated at the end of the long detour that is brought to an end by what we have called *translation 3*. In this phase of the transport, the replication of laboratories, which we have proposed to call the "laboratorization of society," laypersons can once again enter the scene. The world is not always prepared to let itself be absorbed by laboratory science and passively undergo the translations it is offered. In this third type of encounter, what happens between those who arrive breezily to set up their laboratory on new lands to be conquered, and the local people who were there first and usually had not asked them for anything?

To answer this question we will stay with nuclear technology for a while and transport ourselves to the environs of the Sellafield reprocessing plant in England. The sociologist Brian Wynne recounts how, at the start of the 1970s, local people notice that there seems to be an excessive number of infantile leukemias near the plant.[13] The experts consulted reassure the population that nothing abnormal is happening at Sellafield. However, the inhabitants of the area are not convinced; they are sure that something bizarre is happening to them. They therefore decide to organize epidemiological studies themselves, the results of which are given wide media coverage one fine day in 1983 when the BBC broadcasts a program that demonstrates the seriousness of the observations made by the residents. The program highlights the tissue of lies and dissimulations in which the responsible officials were enmeshed. Finally, it is decided to hold an official inquiry. This confirms the excessive number of leukemias without attributing them to a particular cause. In this subsequent rewritten history, the role of laypersons in the identification and formulation of the problem is simply erased. It is decided that everything began with the official inquiry. There is nothing new in comparison with Woburn, except—but the point is important—this expulsion of the residents, who are thus dispossessed of a history in which they were involved from the start. Secluded research is so allergic to interference in general, and to the intrusions of non-experts in particular, that it does not hesitate to hide their contribution! One thinks of those photographs altered by the Stalinists in such a way as to make those who had been physically liquidated disappear in effigy. The idea and the paternity of the inquiry will thus be attributed to Sir Douglas Black, a very acceptable figure in every respect. Having established the facts, it only remains to find their cause. Let the population be reassured; some experts in white coats are dealing with their problems. It is almost a perfect crime. The concerned groups seem to be definitively expelled from a history that is nonetheless their own.

But the residents do not stop thinking, expressing themselves, issuing judgments, or having emotions just because they have been silenced in the public space. A catastrophe, an incident, is enough for what was thought to have been gotten rid of to resurface in broad daylight. The officials thought that confidence had been restored, whereas it was only the right to speak that had been taken from the local people. It was thought that the populations were reassured, whereas they had merely been silenced. Chase away the laypersons and they are back like a shot. The incident which brings the non-specialists back into play occurs in 1986. A moment ago we recalled that 1986 was the year of Chernobyl. The radioactive cloud did not spare England. It led the British authorities to regulate the marketing of meat from sheep raised in Cumbria. Another history begins which is not unrelated to the previous one, and which will pit research in the wild against secluded research. It provides a prime terrain for whoever wants to understand the difficulties awaiting experts in their endeavor to "laboratorize" the world.

The decision of the Ministry of Agriculture to ban the sale of sheep calls into question the fragile economy of the regions concerned. The fears and concerns are broadly alleviated when representatives of the Ministry announce that the ban will last only three weeks. The decision is based on the views of experts who reckon that the source of the contamination, radioactive cesium, should disappear from the environment and the sheep's bodies after 20 days. But the good news is short-lived. July 1986: an extension of the ban is announced, for observations demonstrate that contamination has not disappeared and shows no sign of diminishing. It is then decided that the sheep will be sold, not to be slaughtered, but to be transferred, after being duly marked, to other pastures where they will be decontaminated. The experts continue to exhort the sheep farmers, urging them to stand firm, even if they lose some money, for, they say, it will only be for a time.

It quickly becomes clear that the decision to impose a three-week ban was based on a serious error of the scientists. But this will be revealed only after several months of debate and complementary research. The experts' prediction ("Wait twenty days and the contamination will disappear") corresponded to earlier observations made on an alkaline soil. The experts seriously underestimated the particular, local character of the Cumbrian hills. The cesium, which disappeared elsewhere, remains active and mobile here. Knowledge that was thought to be transposable, because produced according to the canons of laboratory research, proves to be particular and not applicable elsewhere. The Cumbrian grassland falls outside the framework constructed by the experts. The real world is always more complex and varied than the one represented in laboratory models. In the successive translations some variables have been lost, some of which turn out to be of secondary importance while others are revealed to be crucial at the moment of return. This is what happened with the geology of Cumbria.

But an overflow never occurs as a single case. The grasslands where contamination remains are, as we have said, in the neighborhood of the

Sellafield reprocessing plant. The residents, unlike the experts, do not have a short memory. They were expunged from the official history written by Sir Douglas Black! But history catches up with the learned lord. The laypersons reappear in the group photo. There's always someone more specialist! The case is re-opened. Several people then begin to raise the question of long-term contamination from the plant and hidden by the experts. This would explain two sets of phenomena at the same time: the officially established excess number of cases of leukemia and the fact that this contamination that does not conform to scientific predictions. The hypothesis is not at all unreasonable since a serious fire had devastated the Windscale plant on the same site in 1957. Some of the sheep farmers (and who could simply dismiss their hypotheses out of hand?) begin to say that Chernobyl is a pretext, a false cause. Should not everything be imputed to this fire? Let's admit that the theory is coherent. The experts sense it moreover. Without any trace of hesitation, they answer that it is easy to determine whether the cesium, which continues to be radioactive, comes from Chernobyl or the fire by measuring the relationship between isotopes whose lifespan is different. Measurements are made and the experts' judgment comes down, dry and certain like a decision of justice: Without any possible doubt, the radioactivity is due to Chernobyl.

Despite the scientists' fine self-assurance, and maybe even because of it, the shepherds remain skeptical. First, because the specialists have already been wrong once and it does not seem unreasonable to think they could be wrong again. The sequel proves moreover that their fears were well founded: some months later the experts recognize that the observed radioactivity is half due to Chernobyl and half to what are discreetly called "other sources." Later because a serious analysis would have required data from before 1986. Now, despite the farmers' and their representatives' repeated demands, these data were never supplied, the administration finally acknowledging that they did not exist, implicitly admitting that it had not done its work. This cocktail of arrogant certainty, a background of secrecy, and poor work could only arouse the non-specialists' mistrust. In fact, in the farmers' opinion, the most serious thing is not so much that the experts made mistakes, or even that they botched their work, but clearly that they hid all this behind a self-assurance deriving from their status as scientists or experts. The most serious thing is that they refuse to see that the real world—the world of the shepherds and their sheep, a world of limestone hills in which a nuclear plant catches fire one cloudy night in 1957, and a world over which the Chernobyl cloud passes—is not so simple that it can be contained in the knowledge produced, at a distance, by a secluded laboratory. The farmers are reinforced in this bleak feeling when the helpless experts solicit their help in measuring the rates of radioactivity. The experts organize the campaign of measurement without consulting the shepherds beforehand, latter being seen as mere auxiliaries who are barely able to record data on an instrument. But the shepherds know that the divisions agreed by the scientists do not correspond to the subtle geography of their pastures; the zones are very heterogeneous and cannot be reduced to a single climatic or environmental variable: "They [the experts] do not understand this. They think a farm is a farm and a sheep is a sheep." At another time, weary with the struggle, the researchers suggest to the shepherds that they pasture their flocks on other grasslands that seem less contaminated. There is the same disillusioned reaction from the shepherds: "The experts imagine that you stand at the bottom of the hill, and by waving your handkerchief the sheep will rush up at full speed. I have never heard of a sheep that would take straw for fodder!"

The inability of experts to enter into the fine detail of knowledge necessary for a good understanding of the phenomena is even more striking when it is a question of conducting real experiments. One of these, for example, aimed to measure the effect on sheep of bentonite sprayed on their pasture. The farmers immediately commented that no reliable information could be obtained from these experiments. "Their" sheep were not accustomed to being penned in, the effect of which was to disturb their metabolism and affect their health, whatever happens. After some weeks the experiments are abandoned without the researchers at any time deigning to listen to the shepherds.

In these different episodes, different forms of knowledge come into conflict. The local, multi-dimensional, and variable character of the phenomena eludes the secluded science of the specialists. The latter do not see that the big world overflows their laboratory knowledge on every side, sheep proving to be wild beasts that are difficult to control and contain. Now the sheep farmers, thanks to their own apparatuses of observation and memory, have a good knowledge of this world. But the experts are blind to these differences when they take stock of the terrain. They do not see the logic of *translation 3*. At best, they are unaware of the concerned groups, in this case the farmers; at worst, they look down on them, accuse them of irrationality and archaism, and see them as muddled natives caught up in strange beliefs or representations of the world. The researchers think that a sheep is a sheep; the farmers know that such a tautology is a big mistake. The possibility of cooperation between research in the wild and secluded research is lodged in this small divergence. It is because the specialists do not see it that they come up against an obstacle that they cannot overcome.

A conflict of identities is also played out through this conflict over different kinds of knowledge. Shut away in their laboratories, with their data collection and processing schedules, the scientists quite simply ignore the concerned groups, first by erasing them, silencing them, and then by not listening to them when they speak. They reduce a group, with its experience, knowledge, practices, methods of investigations, and ways of living in its environment, to non-existence. They deny the identity of these groups, everything that makes up their richness, their sense of existing and of being caught up in a world in which they have a place.

On the Necessary Cooperation between Secluded Research and Research in the Wild

This is a long detour. Did we have to drag the reader on a guided tour of secluded research, immerse him in the world of patients' associations, take him across continents, and get him to share the trials of angry shepherds or children with leukemia, just to establish the possibility of lay involvement?

It would have been difficult to be briefer. Science is made up of meanders, detours, standing back, and deviations. When science withdraws from the world, it is so that it can reconsider it better. It was crucial not to lose science on the way. We had to stay close to the researchers without for a single moment losing sight of both their point of departure and their point of arrival. The existence of a radical break between secluded knowledge and knowledge in the wild is so rooted in our minds that it was important to follow its fabrication. What does this inquiry give us?

First, it shows us that secluded research gets a good part of its power from its ability to isolate itself, take its distance, and carry out the movement that we have called *translation 1*, which makes possible a realistic reduction of the world. If this is done well, it also makes possible the successful return—what we have called *translation 3*. The force of this revolving movement is that it makes profound reconfigurations of the world conceivable. These reconfigurations, tested in the course of *translation 2*, contribute to the emergence of collectives which were previously improbable, and even unthinkable. But secluded research is not exempt from weaknesses. As a result of distancing itself, it may simply lose contact, cut itself off from the world, and no longer interest anyone.

The risk is greater when we consider disciplines with a long history behind them. Accustomed to living in their entrenched fields, researchers end up with eyes only for the problems which are born in their laboratories. Obviously, the network of connections and translations constructed by secluded research is more complicated than the elementary mesh that we have analyzed and outlined. In an established specialist area there are multiple sources of *translation 1*. Some are situated in existing research collectives, in working laboratories which are constantly producing new research problems; others are external to the constituted research collectives, in what we have called "the big world," and the problems formulated here are less visible and more difficult to reformulate in the language of the laboratory. As these different trajectories intertwine and combine, the real networks may become extremely complex. This is especially so when we add the plurality of outcomes to the plurality of sources. Translation is not a long peaceful river. It is a bit like the Nile, with its multiple sources and its delta, in which the meanders make up a skein which is difficult to untangle. The return to the big world often goes through a succession of stages. From laboratory to laboratory, from one research collective to another, the *Translation* is composed through successive adjustments.

This maze of translations remedies the isolation of secluded research, for it ends up blurring the borders and makes transitions between the world of specialists and that of laypersons. But however abundant this irrigation, however gradual the passage from one world to the other may be, researchers or engineers always end up finding their path closed by a wall separating the territories inhabited by specialists and those in which laypersons frolic. Geologists who scour the countryside to gather data and collect information with the single aim of processing them in the calm of their laboratories are as distant from research in the wild as any biologist or physicist glued to his laboratory work surface. Field work should not be confused with research in the wild! That is why the simplified model preserves all its didactic value. *Translation 1* and *translation 3*, whether composed of a single arch or made up from a multitude of interconnected arches, posit the unavoidable continuity of the movement that leaves the world in order to return to it. Moreover, by hiding themselves behind forests of other laboratories or research collectives, some researchers end up leaving the networks, passing from problem to problem, carried by the wind, without ever leaving the world of secluded research, leaving it to others to maintain or establish connections with the world at large as and when they can or want to. (See box 3.1.)

However, the main weakness of secluded research is not this risk of complete isolation, even if this should not be underestimated. For the most part, the main weakness is the great difficulty this science encounters

Box 3.1

A case of extreme seclusion

At CERN, the Mecca of particle physics, the research collective is one of the most restricted and closed imaginable. The experimentation strictly speaking—in which the accelerator produces and disintegrates particles in order to transform them into other particles and gradually reconstruct the elementary building blocks of matter—only lasts for some months, while the whole experiment may extend over 20 years. First of all detectors must be devised and constructed, which is both theoretical and practical work, and then the campaign must be got under way of gathering data supplied in the form of recording by the detectors, which must then be analyzed and interpreted in order to give support to this or that theoretical option. The originality of the experimentations lies in the abstract character and abundance of the data. Physicists speak of signals to describe the information captured and transmitted by the detectors. Now the accelerator produces a colossal number of signals. The main problem posed by this avalanche is to separate those that are significant from those that are not, to distinguish those that can be imputed to the particles being studied and those that are entirely produced by the functioning of the machine, a bit like the crackling of a radio set sometimes makes the journalist's voice with which it is mixed inaudible. The physicists say that distinguishing the signal from the background noise is like looking for a needle in a haystack. In the hunt organized at the beginning of the 1980s to get hold of one of these particles, the proportion of events (signals) deemed to be significant, and so retained, to the totality of the data recorded was 1 to 10^{10}. A precise knowledge of the functioning of the detector, of its limits and biases, is indispensable in tracking down the "good" signals and identifying sources of error and uncertainty. As no direct measurement is possible, the only strategy open to the experimenter is to go through every possible and imaginable source of error, one by one, and to carry out a constantly updated inventory of all the signals that must be considered as noise. This is a strange catalog, not of what is known with certainty, but of the errors that are assuredly known to be errors! The anti-catalog enables the information to be corrected and rectified: at this point the detectors are idiosyncratic, singular, and not comparable with other detectors, and the events are so numerous and contingent that the constant work of correction requires calculations whose content and results change according to what theoretical bases are called upon. The measurement does not decide between hypotheses or at least make their discussion possible; by means of the calculation of error, the measurement is itself included in the theory adopted. This is a strange research collective, entirely absorbed by the instrument it has devised and constructed to deliver evanescent signals that are difficult to perceive and drowned in a deafening background noise! This is a strange research collective, navigating in a thick fog, knowing that

Box 3.1
(continued)

reality is definitively inaccessible, with the sole ambition of developing a positive knowledge of what are not particles so as to have a better knowledge of what they could be! Karin Knorr Cetina, from whom we have taken these observations (*Epistemic Cultures: How the Sciences Make Knowledge*, Harvard University Press, 1999), notes with amusement that particle physics follows the same route as apophatic theology, which prescribed studying God from the perspective of what he is not rather than what he is, on the grounds that one could not produce any positive assertion about his essence.[1] High-energy physicists thus arrive at the ultimate point of the logic of seclusion. Their one obsession, which we might want to describe as unhealthy if, despite everything, it were not also productive, is to eliminate background noise and expel those parasites and interferences that Cassini feared. Cassini wanted to protect himself from the importunate and intrusive who dared to push at the door of the laboratory; those who for a long time have governed the problem of frontiers and their protection now mistrust their machines, which interfere with the data that they are nonetheless supposed to produce! By a sort of unexpected invagination, the laboratory turns round on itself like a glove: the outside, the source of interference and impurities, is found inside, within the instruments.

Coulomb, with his strategy of burial, is beaten hands down. The simple presence of the machine, however unavoidable, disturbs the experimentation. Like Yahweh's eye following Cain into the grave, the *fureur* of the world follows the physicists into their detectors. The more they distance themselves, the more they live in a universe cut off from everything. *Translation 1* is taken to its extreme point. Nature is made so artificial that it becomes scarcely distinguishable from the artifacts produced by the machine; subjected to unheard of trials, this nature delivers signals under torture which are even more difficult to decode than the Pythian oracles at Delphi. That is why, not content with mistrusting their instruments, the researchers who inhabit this extremely restricted collective, this micro-society, also mistrust their bodies and the illusions of their senses. They constantly multiply and perfect techniques to make the reading and analysis of the traces and inscriptions delivered by the detectors automatic. This extreme point of distance attained by high-energy physics could even prove to be a point of no return. How can you rediscover the world at large when you have done everything to cut yourself off from it? When you mistrust everything, including your own machines and your own body, how can you get back the confidence of those from whom you took your leave? It could be that, by dint of reducing the world by means of increasingly powerful apparatuses, the world has ended up disappearing from the horizon, like sauces that disappear when one reduces them too much. How will physicists find the way back? Is *Translation 3* still possible? Only the future can say.

when it is a question of reducing the world and then of reconfiguring it. Even a laboratory well hitched up to the world, and even researchers who are convinced that they have properly carried out the translations that will enable them to work effectively, will confront insurmountable problems if they refuse to compromise and cooperate with laypersons. We have seen three occasions when there is likely to be conflict or lack of understanding. First, when problems raised by concerned groups do not hold the attention of the specialists, or when those specialists reformulate the questions in an unacceptable way; then when the research collective closes around itself and limits, indeed, in some cases, prevents any debate on the objects and methods of research; and finally when secluded knowledge fails to absorb the richness and complexity of the world, making the simple transportation of laboratories impossible. These are three sources of difficulties that are equally three possible points of entry for laypersons and three possibilities of cooperation, in the dynamic of the production and dissemination of knowledge.

This cooperation between laypersons and specialists is even more inevitable and fruitful the closer we get to domains affecting health or the environment, or, in a word, domains in which knowledge to be produced, in one way or another, concerns the human person in his or her totality. From this point of view, the different disciplines are not all in the same boat. The traditional objects of physics and chemistry are, as such, by construction, external to the human body. The latter may be involved through certain produced effects, but it is not directly involved in the formulation of problems or processes of laboratory replication. Physicists or chemists may therefore dream of reconfigurations of the world that would be obtained by the simple addition of new objects or entities. The *Translation* carried out by secluded research extends the world by introducing new technical artifacts or previously invisible natural entities, along with all the adaptations and safety precautions for their use; it does not drastically change the world.

This dream, which many scientists have shared for a long time, has been damaged, first of all within these disciplines themselves, which have produced numerous overflows. Technical artifacts that were thought to be inoffensive are beginning to threaten people, just as some chemical substances, whose innocence seemed to have been established, turn out to be dangers to health. This is translated into a dramatic return of laypersons who want to have their say and somehow take part in research, that is to say, in the production and practical implementation of knowledge. This desire is asserted all the more, and becomes especially demanding, when secluded research abandons the patiently purified objects of physics and chemistry and invests the new terrains of health and the environment, in which it is increasingly difficult to establish an a priori division between secluded research and research in the wild. The general trend is one of necessary collaboration between the two forms of investigation at the point of the formulation of problems, of the constitution of the research collective, and of the final transposition.

More Than a Specialist, a Specialist and a Half

The reader will have grasped that the aim of this chapter has not been to discredit secluded research; its effectiveness is obvious, and we have not failed to emphasize this. Rather, our aim has been to suggest the possible enrichment of secluded research by showing that it encounters increasingly manifest limits, and that these limits can be overcome only if we acknowledge what research in the wild is capable of.

Why speak of "research in the wild"? Why not be content with a more vague expression, such as "lay knowledge"?

In emphasizing that laypersons are full-fledged researchers in their own right, we are restoring a symmetry that is denied by the usual distinctions between learned and common knowledge, but without confusing one with the other. We will suggest that the model of *Translation* enables us to understand the divergence that separates them and at the same time makes their possible complementarity intelligible.

We can no longer count the coupled notions that have been put forward to account for the cut between the supposed ways of thinking and reasoning of scientists and those of common mortals. We would need more than the great gallery of the Natural History Museum, with its long procession of species that have marked evolution. As examples, we will just cite the contrast between pre-logical and logical thought in Lévy-Bruhl, between *doxa* and *épistemè* in Plato, between savage and scholarly thought in Lévi-Strauss, between pre-scientific and scientific thought in Bachelard and others (e.g., Althusser), or, more generally, between belief, superstition, magical thought, and positive knowledge. Bachelard summarizes this cut with his usual clarity: "To gain access to science is to rejuvenate oneself; the mind is never young, it has the age of its prejudices." Let's make a clean slate of the past—such is the slogan that must be followed literally if we wish to abandon the obscurantism, routine, and ready-made ideas whose only virtue is to facilitate daily life. Opinion, that jumble of prejudices that each takes in without thinking, thinks badly. Indeed, it does not think; rather, it translates needs into knowledge. And Bachelard adds, as a sting in the tail: "In designating objects by their utility, it abstains from knowing them." You cannot be clearer or, dare we add, more contemptuous.

Bachelard was not the first, and surely not the last, to lay into the *vulgum pecus* in this way. Long before him, Plato opened the way.[14] At the end of a brilliant demonstration, Socrates questions Glaucon: "Is it clear to you now that opinion (*doxa*) is something more obscure than knowledge (*épistemè*) but clearer than non-knowledge?" Glaucon, like a good student, replies "Very clear, truly," thus inaugurating 2,500 years of a great division. The hierarchy, which will be transmitted down the centuries, is in place. Up above is scientific knowledge, which goes to the root of things; down below is non-knowledge, which skims over things without fixing attention on them; in the middle is opinion, which is interested in the functionality of things, their utility, in a word, in their appearance. It is a classification perfectly adapted to the luminous metaphor, with the world's opacity on one side and the bright light projected by scientific knowledge on the other. To be worthy of being described as scientific, thought must conform with the requirement that it break with opinion, with that form of knowledge that lets itself be invaded, submerged, and blinded by the world. This distancing, this rupture, this cut that epistemologists want always to be sharper, is also the criterion Popper uses to distinguish the wheat of science from the chaff of non-science. For Popper, scientific knowledge results from a veritable ascesis, an ethic. Science is not produced by a will to know, but by the obsession with putting forward conjectures. The latter are not self-evident, and so they distance themselves from common sense and are intended to be put to the test or, to use Popper's term, to be refuted. Truth, as objective to be attained, is a poor compass, for it is synonymous with easiness: to produce truths is the most banal and easiest thing in the world. An infinitely more courageous and fruitful program is to submit original hypotheses to experimentation in order to probe their realism.

This ascesis, the main purpose of which is to establish a clear cut with common sense, must clearly be fostered and encouraged by institutions in order to be maintained. It nevertheless constitutes an intellectual attitude and a moral choice. Taking leave of the world, tearing oneself free from opinion, accepting the risk of error rather than the comfort of easy certainties, and keeping at arm's length the interests that are supposed to contaminate scientific knowledge, indeed make it impossible, are the values to which the scientist must subscribe. In this way two exigencies defining the conditions of possibility of research are joined together: no science without a cut, and no science without an ascesis of the mind, without disinterestedness. From their creation, all the Academies of Sciences will keep watch over this, and with increasing vigilance. In France, for example, where science is raised to the status of a supreme value, the condemnation of Mesmer and his magnetic cures is a turning point. Lavoisier's decision, which comes like a bombshell, is well known: "The experiments [organized by the Academy] are uniform and also decisive; they allow us to conclude that the imagination is the true cause of the effects attributed to magnetism."[15] However satisfied women may be with their cure, what they think about it is reduced once and for all to the rank of simple opinion; they are the victims of illusions from which only scientific method, as practiced by professionals, can free them.

But this disinterested distancing, which defines the conditions of scientific thought, does not provide the latter with its food. Scientific thought needs information to live on, material to be formed. Here again, Bachelard's response is luminous: "For a scientific mind, all knowledge is an answer to a question. If there is no question, there cannot be any scientific knowledge. Nothing is self-evident. Nothing is given. Everything is constructed." The famous data of experience are clearly never given, they are fabricated, "made" in the framework of experiments devised by researchers in order to answer questions. All the philosophical traditions are in agreement that scientific knowledge is the result of a game of questions and answers. Popper talks of problems to be resolved. The pragmatist tradition, which is so lively today after a long eclipse, says no different.

The Identikit picture of the scientist, or more precisely the requirements to which the philosopher asks him to correspond, begins to take shape. He is a subject who takes his distance from opinion; he tears himself from the world so as to relinquish the interests that could contaminate the knowledge he produces; he devises experiments in order to produce data that feed his reflection and enable him to answer the questions he asks or that are put to him. Everything distinguishes our man of science from the layman, his perfect antithesis. The common man is in the midst of the world, caught in its grasp, unable to tear himself free from the interests that surround him; he is condemned to produce only practical knowledge useful for controlling his daily environment; he is overwhelmed by the sensations that overcome him and toward which he is unable to adopt a critical stance; he reflects them more than he reflects on them, giving them form and classifying them according to categories for which practical effectiveness is the only thing that matters. The common man does not take any distance; he is in the grip of routines, submerged in everyday life.

With these two familiar images before of our eyes, one the negative of the other, we can only hear the powerful voice of Nietzsche: "Take off your mask, mister philosopher! You talk to us of purity and deceive us with words like disinterestedness and moral duty! What, don't you see that science is a diabolical invention which conceals a very different enterprise? Contrary to what you maintain, scientific knowledge is a concentrate of drives, fears, and the will to appropriate. You tell us it is pure and independent whereas it is always dependent and interested, not in itself, but in everything that satisfies the instincts and the institutions that subjugate it." We have come to believe in the Identikit, but of what is it the true portrait? Nietzsche confuses us. What served to define the common man and, by contrast, enhance the image of the scientist, now serves to describe the latter. Nowhere is there a character more obtuse, more in the grip of his interests, or more bogged down in the world than this brave scientist who presents the additional unpardonable defect of being deadly dull.

Caricatures enable features to be picked out and highlight what matters and makes sense. Their defect is that they end up forming a screen which gradually leads us to forget that they are merely puppets, lacking life and depth. Bachelard, Popper, and the others are excessive, extreme, as are Nietzsche and all those who follow him in a symmetrical enterprise of discrediting and relativizing science. The reduction of the scientist to an ascesis, or to a wrenching free from the world, or to being at the mercy of interests that go beyond him, is hardly convincing. But the confrontation has the merit of providing reference points for empirical analysis. Let us start therefore from the model of *Translation* in order to give sense to the different exchanges in the philosophical polemic, with a view to going beyond them.

You have said: taking distance, tearing free, breaking off, cut? Yes, and this is the meaning of *translation 1*. To work at full capacity, the laboratory, as manufacture of knowledge, must be detached from the world, but without however, and this is crucial, breaking loose from its moorings and severing its ties. If the break is to be fruitful, something must be preserved and, consequently, equivalences must be constructed. What takes place in this translation is a change of scale, a realistic reduction which, if the detour is successful, will later enable one to return into the world.

Disinterestedness? Yes, in *translation 1*, which cuts itself off from the world, and in *translation 2*, which is submerged in the research collective; but not in *translation 3*, which returns to it. To understand this strange movement we need only go back to the etymology. *A* attracts *B*'s interest if he places himself between *B* and all the *C*s that strive to seduce and ally themselves with *B*. When a laboratory (*A*) interests the Minister for the Environment (*B*) by "selling" him its project on fuel cells, it must detach the minister from all the pressure groups (C) that try to draw him into other programs incompatible with that of the cells. Interesting B in A passes by way of withdrawing B's interest in C. To reduce the macrocosm, which is essentially what *translation 1* does, we must untangle the ball of existing interests and interrupt all the lines going from B to the innumerable Cs attached to him. But if *translation 1* does not reconstitute networks of interests, *translation 3* will end in failure. The reconfiguration of the macrocosm, which results from the three translations, thus mixes *dis-interessement* and *interessement*. As we have seen in the case of the Institut Pasteur and its diphtheria serum, or of the genetic service and its prenatal diagnoses, attracting interest does not mean following existing interests but working on the list of actors and identities in order to redefine them so that the facts and apparatuses leaving the laboratory find their place and their connections in the new world. World 2 is not deduced directly from the interests in world 1, since it is the result of a detour. In reducing history to the faltering steps of interest, Nietzsche committed an unforgivable logical error. Yes, this socio-technical history, which gives rise to the electric car driven by fuel cells as well as the genetic diagnostic kit, is a somber history of interests, but it is a history in which disinterestedness, as action aiming to withdraw interest, is central. Symmetrically, by denying the work of *interessement* Bachelard sinned through idealism, for without this the laboratory is definitively cut off from the world. Disentangle, yes, but all the better to re-entangle.

Construction of facts? Fabrication of data? Obviously. We can never be convinced enough of Bachelard's and Popper's warning against the naive empiricism that asserts that facts are there to be discovered by the shrewd and visually acute scientist, a bit like the clever prospector who discovers in his sifter the nugget of gold which was overlooked by the non-specialist. But beware of excess! We do not deduce from the fabricated character of facts that any fact whatsoever can be produced. The metaphor of public works is useful to understand this point. The bridges that span the Seine in Paris display a variety of forms and styles that demonstrate to the dazzled tourist that there is not just one way to build a bridge. But what fool would maintain that any form whatsoever and any material whatsoever would do? The man, not in the street, but the man of the bridges, or rather the man who takes the bridge to cross the Seine, knows full well not to venture on any bridge whatsoever! The same goes for the facts fabricated in the laboratories and for the bridges over the Seine. They are constructed, and consequently there are no brute facts any more than there are bridges in the wild state. And if the facts are as varied as the laboratories and experimentations that they organize, they nevertheless cannot be manipulated and fashioned at will. There are facts that crumble, just as there are beautiful and elegant bridges that cannot support their own weight or resist the force of the wind. The laboratory produces an artificial nature, but, as we have emphasized at length, this artificial nature really is real. The soundness and solidity of the facts are linked to material chains of inscriptions produced by instruments, and to their discussion in research collectives. The facts fabricated in research collectives are constructed and real because they are well constructed.

The model of *Translation* shows how taking distance and proximity, fabrication and realism, disinterestedness and *interessement* are the two sides of the same coin, or components, which cannot be disentangled, of a single process of the production and application of scientific knowledge. At the same time the model frees us from the seemingly insurmountable opposition between specialists and laypersons, between scientific thought and common thought, while explaining the interest of their distinction. It enables us to understand how actors who are not professional researchers can nevertheless be integrated within the dynamic of research. To speak of research in the wild is to emphasize a form of involvement in which what counts above all is the formulation of problems, the modalities of application of knowledge and know-how produced, as well as the necessary opening up of the research collective. In short, the model prefers the concept of research to that of science.

Whether professional researchers or researchers in the wild are concerned, the starting point is constituted by problems deemed to be bizarre, incongruous, disturbing, and unexpected; in both cases the objective is to dissolve irregularities into regularities, into causal chains; in both cases experiments are organized, observations are made, things are learned; in both cases there is the requirement of open discussion ensuring the widest possible confrontation of different points of view and interpretations; in both cases the question of implementation, of the transformation of the world, is tackled. The only difference, but it is a major one, is in the way in which the different moments of the *Translation* are prioritized and organized. Researchers in the wild see the seclusion of research as a simple detour which should not conceal the importance of problematization and return. They exercise their vigilance when the secluded researchers are engaged in *translation 1*, in order to be sure that the problem translated is their problem; they follow the work of the research collective at the point of *translation 2*; and they are attentive to the course of *translation 3*, when the answers to their supposed questions are passed on to them. And if researchers in the wild develop this sensibility, it is because what is at stake in the *Translation* is their identity and existence.

These two modalities of research are adapted to each other, they are made for cooperation. We have suggested, dare we say shown, that research in the wild is perfectly compatible with secluded research and ready to collaborate with it. Complementarity, mutual enrichment, and not opposition. Connecting up the two forms of research enables the advantages of each to be combined, while erasing their respective weaknesses. Research in the wild brings with it a tremendous force, that of a collective—sometimes in the process of being formed—which is identified with the problems posed and extraordinarily active in the implementation of solutions. Secluded research supplies its strike force. The detour it organizes opens the field to the most improbable experiments and translations and, as a consequence, to a more open range of reconfigurations of the collective. Put more simply, specialized research is vascularized by lay research. Alternatively: Without ever ceasing to exist, the research collective is constantly plunged back into the social world from which it came. In this way the three confrontations we have described are reduced, as we say a fracture is reduced, at the very moment they occur.

The modalities of cooperation between secluded research and research in the wild are clearly very varied and to a large extent remain to be invented. Chapter 5 will provide some indications on the procedures that foster this cooperation. But whatever the modalities may be, what is at issue in the establishment of different forms of cooperation is the invention and organization of what should be called a collective investigation and experimentation that involves constant to and fro between specialists and (concerned) laypersons.

Collective experimentation develops along two closely intertwined dimensions. As the different examples presented in this chapter have abundantly shown, it is in fact difficult, indeed impossible, to distinguish the production of knowledge strictly speaking from the production of social identities. What is involved when the inhabitants of Woburn or the parents of children with AFM battle to get their problems recognized is really the recognition of their existence and the legitimacy of the difficulties they have to resolve. Failure to accept the questions they raise is to consign the first to an absurd misfortune and to oblige the second to hide what they are told is the defect of their children. When those with AIDS reject the experimental protocols used for clinical trials they are struggling for the recognition of scorned minority identities. When English shepherds refuse the verdict of the experts and throw themselves into a recomposition and adaptation of knowledge, they are raising their voices in order to get recognition for their threatened identity as sheep farmers. It would be tragic to separate these two dimensions and say, for example, that only wounded identities are involved, confining laypersons to the level of emotion and passion, or only to see them as informants, precious auxiliaries of laboratory science. Science and passion, knowledge and identities are inseparable and co-evolving. They nourish each other. That is why science and politics go hand in hand. That is why the procedures to be devised to organize this collective learning, all of which are directed toward the constitution of a common world, must allow for the simultaneous management of *both* the process of the fabrication of identities *and* the process of the fabrication and incorporation of knowledge. We have a better understanding of why researchers in the wild demand to be heard and to be associated with the *Translation*.

4　In Search of a Common World

If we take the term in its strict sense, there never has been a real democracy, and there never will be.

—J.-J. Rousseau, *The Social Contract*

Tanned as if they have just spent some weeks relaxing in the sun, some engineers of the CEA (Commissariat à l'Énergie Atomique—the French Atomic Energy Commission), who have been working relentlessly for several years to make their research program credible, are taking a well-earned rest. The research program is under close watch, and day by day it comes up against an ever-growing number of increasingly fierce adversaries.

What is to be done with nuclear waste? This is the simple question for which they strive to find an answer. But not *any* answer, and that is why they have agreed to these three days of continued training whose organization has been entrusted to us. They know that times have changed, that we have bid a definitive farewell to the blessed years of technicist euphoria, so well expressed in the often-cited motto of the Chicago's 1933 world's fair: "Science finds, industry applies, man conforms." They have learned that nothing is ever so simple and that it often turns out that society refuses to follow. These and other engineers have found a name for this strange insubordination that is translated into an incomprehensible and irrational rejection of progress. In their jargon they call this the "social acceptability" of technology.

Nuclear waste is an exemplary case for those interested in this disease, its clinical picture, and the possible treatments that should enable us to overcome it. For many it is not too much to say that French society is sick of its nuclear waste, as we say that some people suffer from digestion disorders. For dozens of years, these engineers and their elders have developed a nuclear industry that they deem to be for the common good. They have done this in a secret and hidden way, in the mysteries of their offices. Oh, they were not just purely and simply defending particular interests disguised as the general interest. They were not in pursuit of their personal enrichment. Certainly, once in place, industrial and technological programs end up secreting interests that ask for just one thing: that the programs continue and become increasingly irreversible so that there is no force strong enough to challenge them. But it would be unfair to say that these decision makers, these senior civil servants, graduates of the Republic's elite engineering schools, having a strong sense of solidarity with the body to which they belong, were driven only by the lure of gain or the intoxication of power. Their sin is serious enough without charging them with ones they have not committed. They have merely wanted the people's happiness, without letting them say a single word and without inviting them to sit around a table to discuss and negotiate. And if they have ignored them or silenced them, it is not because they are in principle enemies of democracy. Rather, it is because they want the people's good that, with aching heart and after an intense effort of intellectual exertion, they feel obliged not to listen to them: not the least of the people's defects is that they do not know what is good for them. The people are the dark and primitive Middle Ages of primary-school history books. And the men of the professional corps of engineers are the light guiding the people. To plot the way and go as quickly as possible they are not frightened of being a bit cynical, a shade Machiavellian, and of resorting to dissimulation. Furthermore, they do not hesitate to let the cat out of the bag when a visitor from the United States asks what seems to them an incredibly naive question: "I wanted (I explained) Monsieur le directeur to tell me about the scientific and technical decisions in which Monsieur le directeur had taken part during the 1950s and the 1960s. Imagine my surprise when he slapped his hand on the desk, leaned toward me, glaring, and roared: 'But Mademoiselle! These were not scientific or technical decisions! They were economic decisions! Political decisions!'"[1]

Nuclear energy, they say, is like abolition of the death penalty. Every educated person knows that morality and reason require us to be in favor of them. Every educated person also knows that the uncontrollable crowd can only be opposed to them, for the people follows its instincts and allows itself to be lead astray by its passions. The fault of these decision makers is this aristocratic belief and not the defense of their particular interests or an obsession with power. Or at any rate, it is not solely this. The fault is thinking that democracy can function only if the people are kept at arm's length.

Every fault deserves a punishment proportionate to its gravity. Who has sinned against democracy may pay with an increase in democracy. The eleventh commandment: anyone who silences those who should speak is condemned to organize ways for them to express their views! Those who, thinking to do good, have transformed nuclear energy into an exclusive reserve, are now forced to open the doors and windows, put the files on the table, and abandon bypassing strategies and the bribing of elected representatives. If the end justifies the means, only debate can justify the end. Radioactive waste has become socio-active. We thought we could get rid of it once and for all by burying it deep in the most inert clay or the most compact granite, protected by thick containers. This failed to take residents and future generations into account. Some high officials, some of the bolder ones, even dare to suggest that it is a question of the return of the repressed. According to them, the technocrats have been caught out by their decisions.

Social mobilization and the so-called Bataille law of 1991 opened up the game and provided a framework for discussion, making debate possible, indeed necessary. Questions that had been suppressed returned in force. What do we need to know to manage radioactive waste in the best way? Who is concerned by this decision, and on what grounds? How can the debate be organized so as to prepare the measures to be taken? All the engineers now know that the *demos*, the people, are back with us, and that these questions are now in the public space. The people had been dismissed; now they are back at the negotiating table. One will have to live with them, listen to them, even if it is only to be forgiven for the sins committed in the past by colleagues who were a little too arrogant, a little too sure of themselves.

Like all good engineers, those seated around the table feel that the strange constraint imposed on them—talking with the people instead of speaking in their place—could be turned into a strategic resource. The Bataille law has in fact provided for the exploration of three options: transmutation, deep burial, and surface storage. We are in France, where we like things to be clear. Each of these options has been assigned to specific teams of engineers, which have been instructed to avoid duplication. Those present today have inherited the third option, which seems the least plausible, the least realistic. Is it really reasonable to store above ground nuclear waste that has a lifespan of hundreds of thousands of years? This intermediary position appeared in 1991 as a real non-option, at best a transitional stage pending definitive transmutation or irreversible deep burial. One of the first consequences of putting the issue of nuclear waste up for debate is to force a revision of this hierarchy. What the non-specialists fear are irreversible, irrevocable decisions, that is to say, roughly, deep geological burial, because transmutation is still in the realm of dreams.

The engineers who are here today are quite clear that the people reduced to silence are on the side of those working on deep burial, whereas the talkative people, those who are forever giving their opinions, whom one hears on the 8 o'clock television news, who take to the streets, and who pursue the experts to the borders of their departments, are ready to support interim solutions so as not to lose sight of the waste, so as never to repress nuclear energy in the depths of the earth and the collective unconscious. In the intensifying competition between options they thus have an interest in the people speaking, expressing themselves, and being even more talkative. Furthermore, to profit from this advantage and make the preferences of the non-specialists more realistic and solid, they have invented a new option of long-term sub-surface storage, which tends to become an option in its own right and no longer a provisional solution.

It is one thing to be anxious to talk with the people, since one has an interest in doing so, but it is another thing knowing how to set about it. Furthermore, who are the people, this *demos*, this elusive character of every democracy that everyone talks about but no one ever sees? Some want to talk to the public, others to pressure groups, others to citizens, and others to users and electors. And what does "talking" mean? Do we really have to organize a bone fide dialogue? Do we have to pretend—and even so, the affront is difficult to swallow—that these laypersons, these non-specialists, are able to talk about technical matters?

Once the decision has been made to open up and come out from the mysteries of power, everything still remains to be done. In the present case, the first move is to turn again to the social sciences, for they have good and loyal service records. During the decades of the great silence social scientists actually played a discreet but essential role. How can we silence the people and speak in their place? This was the question put to them, and it so resembled what they were accustomed to doing that they did not hesitate for a moment to help the technocrats in difficulty. In comparison with traditional forms of know-how, the social and human sciences actually wield a terrible power: that of discrediting actors' words. Just as it is difficult to prevent someone from speaking (short of resorting to physical violence), so it is easy to interpret what is said so as to discover beneath the words, beyond immediate significations, a deep, hidden meaning, in short the true signification of the words uttered. Sociologists and anthropologists shout from the rooftops: "The people speak, but it is not really they who are speaking! The people think they are opposed to nuclear energy; they think they are demonstrating against the establishment of La Hague, against Superphoenix; they think they see leukemia in the nearness of the Woburn dump. In reality they are expressing irrational fears, a terror that is endlessly renewed in the face of progress, change, and the disruption of traditional frameworks. Look at their faces: you can read on them the fear of future uncertainty. Listen to those angry voices: you can hear in them the tremors of the fear of change. Forgive them Father for they do not know what they are saying!"

The social sciences currently claim the exorbitant power of restoring a meaning that they fearlessly assert it is precisely their mission to disclose. Called up to the front line, social psychologists do not try to silence the people; it would take battalions of policemen, judges, and social workers to achieve that. They limit themselves to shifting the origin of the discourse, attributing it to irrational anxieties that must be taken into account, not in order to make decisions, but in order to get them accepted.

Once raw, spontaneous speech has been disqualified, it remains to construct civilized, managed speech. The social sciences also know how to do this, for they have invented a whole range of techniques and methods for asking "good" questions which enable one to get "good" answers, like opinion surveys, questionnaires, or ethnographic studies that get closer to the natives. The marvelous thing about the social sciences is that they are able to silence people and to get them to talk at the same time.

Faced with new difficulties, why not call on them once more? Why not ask them, not to silence spontaneous speech so as to replace it with domesticated and reworked speech (somewhat like the fabrication of new melodies with the sound mixers of DJs and sound engineers), but to shed light on the organization of the debate and put forward procedures for dialogue with the people? The engineers are here today because they are convinced that some sociologists can suggest guidelines for this question. And we are here today because we are convinced that the social sciences actually can play this role of participation in the organization of public debate. To provide proof that dialogue is possible and fruitful, we have decided to show a videotaped documentary account of the debates of the French citizen conference on genetically modified organisms that took place at the start of 1998.

The history of this conference is complicated. Since we will talk about citizen conferences at greater length later, it will suffice to say that this one, the first of its kind in France, allowed fifteen ordinary citizens to become informed about a complex issue, and to enter into dialogue with experts and representatives of pressure groups, in order finally to make a series of recommendations to the political decision makers. These decision makers are really in an awkward position: on one side are forces that are pushing them to authorize *both* the cultivation *and* the import of transgenic plants such as maize and soy; on the other side, non-governmental organizations (NGOs), consumer associations, labor unions, and political movements are fiercely opposed to this and demand a moratorium. It is a hot subject on which the decision makers are undecided. The experts are divided, and it is difficult to identify and stabilize the social forces, each new decision provoking the emergence of actors who were previously silent. Will this panel of fifteen ordinary citizens be able to clarify the debate? Will it not help, rather, to make it even more obscure and unmanageable?

If these questions were to be put out of the blue to the CEA engineers undergoing training, despite their interested openness they would no doubt be inclined to answer the first question negatively and the second positively. On such complicated issues, what can you expect from non-specialists, from housewives over 50 years old from the depths of Lorraine, or from farmers still astonished to find themselves in the luxurious rooms of the National Assembly? But, since they are open, they have agreed to suspend judgment, just long enough to re-establish contact with reality, in this late August. They have agreed for the lights to be turned off so that they can watch a video put together by one of our colleagues, summarizing, in little more than 90 minutes, dozens and dozens of hours of discussion, cross-examination, and reflection.

When the lights are turned back on, the room has become silent. These engineers have tough skin; they are used to the most violent and malicious attacks. They are thoroughly familiar with all the arguments for or against nuclear energy, for or against this or that option for the management of nuclear waste. Proof of this is the astonishing role playing to which they submitted on the first day. Each of them had to defend a position: one had to defend the position of the Confédération Générale du Travail (the most powerful French workers' union), another that of angry viticulturists, another had to set out calmly the point of view of the Agence Nationale pour la Gestion Des Déchets Radioactifs (National Agency for Radioactive Waste Management, abbreviated ANDRA), and yet another had to give an account of the government's decisions. We were misled: their exchanges oozed with raw realism, the angry outbursts seemed even more genuine than those to which the media accustoms us. In short, it is impossible to reproach them with not listening to or not understanding what is being said and discussed. They know everything about the subject; they know "their" nuclear energy by heart. They have read all the books and have seen all the television programs; that is why they are gloomy and blasé. And yet here they are dumbfounded! Those who are accustomed to speak in the place of non-specialists—the role playing had brilliantly evoked this— are rendered mute by the words of non-specialists. Finally one of them decides to say what each would have liked to say but could not manage to express: "It is moving." We were ourselves struck by the video, which we were watching for the first time. And it seems that all those involved with the conference felt the same emotion. However the conference itself is judged, what was or was not got out of it, it remains the case that it vividly demonstrates that ordinary citizens can take the floor to say sensible, intelligible, and serious things. And above all that this speech is moving.

What is the reason for this emotion? What is the reason for this strange sentiment which means that we find ourselves concerned and affected by what is said and the way it is said? Everyone's views are solicited again. Maybe these clear-sighted engineers, who are not much given to emotion, know the answer. "What is striking," one of them confesses, "is that they can abstract from their personal interests, adopt the point of view of the general interest, ask good questions, and finally come up with moderate recommendations." "You trust them," another adds. This says it all, or almost. When they listen to a politician or one of their engineer or technocrat colleagues, these researchers, like good psycho-sociologists, immediately hear the discourse of interests, the language of corporatism. They decode the calculations. Even when it is a question of the common good, of the collective interest, they know that it is turnovers, export opportunities, monopoly incomes, or electoral calculations that are at stake. But how can we suspect the farmer, who just the day before was driving his tractor in Flanders, of confusing his preferences with the common good? How can we imagine for a moment that this housewife with the harsh voice is seeking to profit from her judgments on the innocuousness of GMO or the degree of uncertainty of knowledge? The members of the panel are so far from the stakes linked to transgenic plants that they have no difficulty in distancing themselves. They have been so well placed in the position of those who must consider the issue from the point of view of the collective interest that they have no difficulty in adopting this role. They produce that astonishing metamorphosis which seemed self-evident to Jean-Jacques Rousseau: constructing a general will on the basis of particular wills. And the practical recipe suggested by the author of *The Social Contract* is not far from being applied: "If, when the people, being furnished with adequate information, held its deliberations, the citizens had no communication one with another, the grand total of the small differences would always give the general will, and the decision would always be good. But when intrigues arise, and partial associations are formed.... The differences become less numerous and give a less general result."[2] They are only fifteen, but they are so different—and have been selected to be different—and they are so far from the intrigues and leagues of GMO, that they have no difficulty in making this improbable point emerge, this geometrical spot that is so difficult to localize: that of the general interest. They are obsessed, as the video shows, with differences: they gather the points of view of all the possible concerned groups; they worry about farmers, but not only big landowners from the Beauce area, about the economy, jobs, and consumers. They successively adopt the points of view of each, making an effort of imagination to explore all the possible overflows so that they can identify all the groups involved, and in the process they consider the kind of knowledge that has to be produced in order to arrive at a fuller and more accurate picture. It is by performing this unusual exercise—putting oneself in the place of each and, in order to do this, identifying what the places are—that they manage to bring together such different interests and points of view and find a common position which is clearly provisional. The video shows that this is possible. That certain conditions have to be met for it to be possible is shown by the impasse in which the role playing ended: "It's difficult to see how a solution can be found when you consider the extent of the oppositions and differences," guffawed the engineer who for 40 minutes played the part of an angry viticulturist with disconcerting application and frightening effectiveness.

The citizen conference seems to contradict these gloomy words. It is possible to construct a place that allows access to all the other places without reducing them, to construct a role that allows all the other roles to proliferate without suppressing them. And, thanks to a procedure, these fifteen ordinary citizens play this singular role. The spectator feels affected, moved, because he feels concerned by their discourse. Before seeing the video, he was on the outside, having no opinion on the subject, or ready to leave it up to his preferred spokespersons. After seeing the video, he is aware of the diversity and legitimacy of the different points of view, while realizing that it is possible to give a fair and measured account of them, and—why not?—take them into account in the decisions to be made.

Role playing that ends up in the disillusioned observation of an impasse from which no one can extricate themselves because the interests seem to be so entrenched and incompatible. A citizen conference which gives instead the feeling that it is possible to define a place where singular wills are combined and where it is possible to imagine a common world that can accommodate differences which seemed to be irreducible. A public debate that allows us to see that those who work out the general will in this place are laypersons ignorant of everything! To be sure, this is only a very imperfect small-scale model, a staging with obvious limits. But that's not the main thing. Here is the proof that it is possible to give the floor to the people, without fear of this word, and without plunging into the irrational and obscurantism. The people even manage to produce an effect of clarity that the experts, lost in their professional knowledge and interests, fail to produce. Here is demonstrated that what matters are procedures, the procedures alone, the rules or organization of these debates and discussions. The common will is not discovered by chance. Ruthless rules are needed. Without the drawing of lots for a panel of representative and non-concerned citizens, without the training sessions, without the hearings of experts and pressure groups timed almost to the second, role playing would have regained the upper hand, and with it the dialogue of the deaf and the struggle of all against all.

When Ordinary Citizens and Laypersons Challenge the Great Divides

Representation and Consultation: A Question of Procedures

We have known since its origins that democracy is mainly a matter of procedures. Just as we know that democracy is an enterprise that is never completed and consequently that procedures must be constantly evaluated and revised.

The notion of representation is at the heart of these procedures. There is no democracy in which there is not a break between representatives and those represented, and one of the sources of the variety of democratic regimes is the diversity of the forms of organization that lead to the replacement of the people in its entirety with a handful of spokespersons who govern in its name. It is not an imperfect but unavoidable procedure to which we resort for solely practical reasons. It is not because the assembly of all the citizens would be unmanageable, especially owing to its size, and has to be replaced by a smaller assembly. In fact we should resist the idea that the people is made up of individual citizens each of whom knows exactly what he or she wants on every subject and is endowed with preferences that are fixed once and for all.

Such a people does not exist. And if it existed, the problem of its representation would continue to be insoluble, at least on paper. We have known since Condorcet, and Arrow has given a faultless demonstration of this, that even if the people was made up of citizens knowing exactly what they wanted, the work of aggregation and representation by which collective preferences could be deduced from individual preferences would be no less doomed to failure technically. Representation is work that is constantly being taken up and started again, and not a simple objective description; it is founded on the more basic mechanism of consultation. The person represented does not always know what he wants; it is in the debate preceding the choice of his representative, in discussion with him, that he gradually learns what his preferences are and his will is gradually formed. Without representation, viewed as the process in which wills are formed, there would be neither individual will nor common good. Representation and the consultation that underpins it fabricate the person represented together with the one who represents him. The latter says to the former "I say what you say," and as a result the person represented is rendered loquacious. Without spokespersons there is no voice; without debate between the person who speaks on behalf of another and the person on whose behalf another speaks, no speech is possible. Representation is not a second best, an ersatz of direct democracy. It is the cornerstone of democracy, since it is representation that gets the people to speak and at the same time designates their spokespersons. All those who have emphasized the constitutive role of the break between those who are represented and their representatives are a thousand times right. Democracy is in fact inscribed in this ever-open gap. To suppress it would be to deny the very conditions of existence of democracy (since no general will could be calculated); to accept it is to render democracy practically possible but always imperfect (since representation simultaneously produces speech and silence: "I say what you say, so you are invited to remain silent at least for a time and on some subjects"[3]). The general will and individual wills are constructed at the same time: agreement is possible only on this condition, but its price is the at least provisional silence of those represented.

There is nothing natural or spontaneous about this mechanism that makes it possible to get citizens to speak, designate their spokespersons, and, by organizing this delegation, silence those represented. It is necessarily organized. How can consultation be set to music? How, and for how long, are representatives to be designated? How can we enable those who are represented to denounce what they see as their betrayal by the representatives? Procedures play a crucial role in the answers given to these questions. They shape the strange alchemy that gets citizens to speak while silencing them, and which can result in the formulation of a general will only because it refuses the *a priori* existence of individual wills.

Faced with such a volume of responsibilities, procedures can only be approximate and makeshift. Representation is in constant crisis, and above all in those states which are considered to be advanced democracies. Representation never reveals its limits so much as when it has been pushed as far as possible. In countries where representative democracy[4] is scorned it is seen as the most precious good and its least perfect forms are tolerated; where it has been established and has become a horizon that cannot be crossed, its limits are denounced and there are no words harsh enough to condemn the violence it is deemed to be guilty of when it legitimizes the exclusions on which it is based. In a word: those who are deprived of democracy sometimes aspire to it; those who enjoy democracy tend to vilify it or devalue it. The former struggle so that the people may finally be represented; the latter insist that this representation is never perfect enough.

Whether it is the interminable questioning of the regime of political parties, the condemnation of their sclerosis, or the denunciation of the growing breach between the real people and their representatives, the critique of democracy is obviously anchored in the paradox of representation. Since to represent is to silence, and since any practical implementation of representation tends ineluctably to maintain the breach between spokespersons and those who choose them, at least for a time, existing procedures are inevitably challenged and denounced. "Don't you see," it is said from every direction, "that it always the same people who speak; can't you hear the deafening silence of those who are denied a voice, because it has been confiscated from them, and who will never be able to express themselves because they have been deprived of any means of doing so?"

The more democracy, the more representation—such is the logic at the heart of democracy. It is expressed in a formula that sounds like a slogan: democracy must be democratized. It is a slogan that constitutes what the medievalist Alain Boureau calls a collective statement (*énoncé collectif*), which sums up in a few words (like *vox populi vox dei*, which Boureau takes as an example) an aspiration and a belief that everyone shares and that orientates the action of each individual, while leaving to each individual the choice of the precise meaning that is given to the statement.[5] Who would dare to be opposed to the constant deepening of democratic mechanisms? Who could say what precisely this involves? The collective statement, both precise and ambiguous, has the fantastic power of making energies and projects converge without erasing the variety of points of view and conceptions.

If the criticism of representation is both constant and very actual, it is particularly acute in the case of science and technology. Our first chapter and the presentation of what we called "hybrid forums" showed this. In those cases, the general procedures which were developed over time to enable citizens to speak tend to become leaky everywhere. New procedures are devised and desired which will enable the deficiencies of those in force to be overcome. Let's be clear: Hybrid forums do not call democracy into question; they demonstrate and express the need for more democracy, for a deepening of democracy. They are one of the particularly visible and urgent manifestations of the more general movement that calls for the democratization of democracy. The simple fact that they are not purely and simply repressed, even though some established forces try to reduce them to silence or non-existence, and the simple fact that they mobilize opinion, although many interest groups strive to devalue them, demonstrates their legitimacy, if this must be demonstrated. Everyone knows that they are not undermining democratic procedures but are instead entirely set on enriching them. Hybrid forums are therefore precious laboratories. What they obviously express is a criticism of the procedures on which representation is usually based. What they demonstrate in practice is a desire for public debate, a demand that groups which are ignored, excluded, and often reduced to silence, or whose voice is disqualified, have the right to express themselves, to be heard, to be listened to, and to take part in the discussion. The definition of the common world, in which each is called upon to live and means to find their place, cannot be left to spokespersons who are no longer in tune with the moving reality of the *demos*. These new cases overflow the democratic procedures which are common to the political regimes of advanced societies. The socio-technical controversies to which they give rise, and which spread beyond political parties and legitimate authorities, emphasize the need for procedures more open to debate, more welcoming toward emerging groups, and more attentive to the organization of the expression of their views and the discussions it calls for.

How can we devise the enrichment of procedures? How can we devise forms of consultation that do justice to the diversity of points of view and aspirations? Answers to these questions are not to be found in any manual. They are invented, and tested on several fronts by the actors themselves. Science and technology is not the least of these fronts. Hybrid forums are experimentations under real-life conditions that enable the analyst to grasp the limits of existing procedures (since these forums are born from their impotence) and to assess the contributions of those invented by the actors (since the latter devise new forms of representation and consultation in the heat of the action). And it may be, in addition, that the solutions put forward by the hybrid forums can be transposed, carried over into other fields where science and technology are not necessarily central, and thus contribute to the more general movement of the democratization of democracy.

The Development of Hybrid Forums: A Criticism of the Limits of Delegative Democracy

Through their continuous overflows, hybrid forums highlight the difficulties representative democracies face in managing situations of uncertainty. These uncertainties may be grouped into two big families: those concerning our knowledge of the world and those affecting the composition of the collective.

What do we know about the world? How is the collective in which we live made up? Chapter 1 showed us that our democracies block the open exploration of these two questions by introducing two sharp breaks, two big divides, which vary in extent from country to country, but which always reappear when the political stakes of science and technology being debated. We have seen that hybrid forums are more or less spontaneous, more or less organized endeavors, and also, in their diversity, apparatuses, for the trial-and-error exploration of possible answers to these questions surrounded by radical uncertainties.

The first of these breaks is that which leads to the isolation of scientists. This isolation is the result of a delegation by which society entrusts specialists, the scientists, with the task of producing sound forms of knowledge, certified knowledge. Shut away in their laboratories, researchers are accorded complete autonomy, with increasing budgets, but in return, and this is the object of the delegation, they must come back with confirmed facts, as solid as the hardest granite. Autonomy and billions of euros is the price the collective pays these luxury mercenaries whose sole mission is to produce knowledge purged of all uncertainty. "Do what you like in your laboratories, spend as much as you need, but do not come back to see us until you are sure of what you put forward, before you can describe with the greatest certainty all the possible worlds in which we could live!" Nothing is more normal than scientists disagreeing with each other! Nothing is healthier than them being opposed to each other on how to conduct an experiment or interpret its results! Science is made of doubts, trial and error, and divergent interpretations. But its grandeur consists precisely in overcoming them to arrive at a meeting of minds. And the production of the truth, of agreement, can take place only in a closed field, between specialists. They are the ones who must decide on the validity of knowledge. Disorder very quickly gains ground if disagreements are made public. Thing very quickly get out of hand if laypersons are allowed to take part in the discussion of experiments and their results. The main ambition of this first delegation is to avoid the confusion of roles. Above all it aims to ensure that scientists have a monopoly on the production of knowledge. As a result of this, the uncertainties linked to the knowledge produced, which enables possible worlds to be described and brought about, is confined in the laboratories. The only thing to leave the laboratories is certain and pacifying knowledge on which political debate can be developed like a superstructure sure of its bases. Our democracies have not ceased for a moment to play off secluded research against research in the wild, thereby ensuring the separation of political and scientific spheres.

Once politics has been purged of all scientific uncertainty, thanks to the great divide between specialists and laypersons, it remains to organize the debate that should lead to the expression of the general will. This is where the second delegation comes in, which produces the second break: the delegation of elected representatives by ordinary citizens with a view to the constitution of the collective. A crucial role is played here by the electoral ballot in which citizens take part in the election of their representatives at the end of a public debate organized so that they can choose between the different candidates who offer to represent them. This procedure in fact produces five reductions that end up in the second great divide. The first rests on a massive exclusion of all those who are not called to the ballot box and who, as a result, are transformed into outsiders: for some this exclusion is taken for granted and does not pose a problem, for others it is felt as an arbitrary act of violence. The second likens this limited and circumscribed collective to a collection of individuals who are seen as being independent of each other and endowed with an autonomous will and power of judgment: groups, as such, do not have a say in the matter. The third limits each of these individuals' capacity of expression to the choice of one or several candidates from a pre-established list, and indeed, in exceptional cases like a referendum, to a yes or no answer to a simple general question. Through a more or less complex statistical calculation, the fourth replaces the population of citizen electors with a more reduced population of representatives. Finally, for a period determined in advance, the last reduces to (relative) silence those who at the end of this procedure have become the represented, granting those who have become their representatives an almost exclusive monopoly of speech on any political subject whatsoever. This fivefold reduction in the delegation by which an ordinary, individual citizen is constituted, who entrusts a general mandate to his or her representative, hollows out a gulf between this citizen and the spokesperson to whom he or she has delegated the power to decide on the composition of the collective. It may lead to the constitution of a closed universe of professional politicians. The latter, supported by parties which mobilize the strategic resources, compete with each other to capture the votes of the electors and develop programs whose main purpose is to enlarge their electoral market.

What the simple existence of hybrid forums underline is precisely the institutionalized character of these two delegations and the breaks they give rise to, and consequently the difficulty in getting round them. By giving prominence to uncertainties concerning states of the world and the composition of the collective, socio-technical controversies reveal the otherwise invisible mechanisms by which what we have chosen to call delegative democracy usually manages these uncertainties. By delegating the production of knowledge to specialists, who are granted an almost exclusive monopoly moreover, delegative democracy purges political debate of all uncertainty regarding possible states of the world. By constituting itself as a political body made up of individuals (citizens) endowed with a will and definite known preferences, delegatory democracy excludes all uncertainty on the composition of the collective, since the latter is reduced to the aggregation of individual wills which are supposed to be perfectly conscious of themselves.

The symmetry of the procedures on which delegative democracy rests will not have escaped the reader: two massive reductions, two exclusive delegations, and two sharp breaks.[6] The first separates specialists and laypersons[7]; the second carves out the gap between professional politicians and ordinary citizens. The two breaks produce two populations that previously did not exist. In fact it is the very movement of delegation—whether that by which laypersons leave the production of knowledge to specialists, or that by which ordinary citizens entrust their representatives with the task of composing the collective in their name—that leads to the existence of *both* the layperson *and* the ordinary citizen, and with them, as their corollaries, *both* "the" specialist *and* "the" representative. This double removal confines debate on the state of knowledge to professional researchers and debate on the composition of the collective to spokespersons who tend to take over the voice of those they represent. An extraordinary and fruitful invention was needed for this drastic restriction to be possible, and for the people to agree to be silent and without a voice. Since there are uncertainties everywhere, since they undermine collective life from within by allowing the emergence of unexpected groups to remain as a constant threat, and since they make it difficult to foresee and control the events that form the weave of the history of the world in which our own history is mixed, the best stratagem is to create specialized institutions for dealing with them: laboratories for the first, parliaments for the second. A result of this is the replacement of the uncertain *demos* with the individual in the form of that reassuring figure of the layperson who is also the ordinary citizen.

The balance of this set of arrangements is fragile, and it is this fragility that gives it both greatness and legitimacy. It makes democratic delegation rest on a paradox: the silence to which the layperson and the ordinary citizen are reduced, and without which there would be neither ordinary citizen nor layperson, is a silence that is desired, accepted, and contractualized. At any moment, *both* the layperson *and* the citizen, whose reality is affirmed and recognized, can break the silence and become indignant that they, who do not speak, or speak so little, are not listened to. "Be careful, if you, who are nothing without us, persist in ignoring us, we are going to make a row!" To prevent the alarmed cries of the layperson and the indignant cries of the ordinary citizen, to avoid them filling the streets with their boisterous and wordy protests, delegative democracies have, of course, invented numerous outlets.

One way to discourage the untimely voice of an ordinary citizen demonstrating against the censorship of which he feels he is the victim is to multiply electoral consultations and representative agencies. The ordinary citizen thus finds himself being offered ever more numerous, but tightly disciplined and framed occasions to have his say on subjects which are also increasingly varied, but which he does not choose. Moreover, the broadening of consultations may be backed up by the benevolent supervision of spontaneous forms of giving voice (demonstrations of all sorts) that permit the organization of controlled overflows in relation to the electoral apparatus *stricto sensu*. In both cases, the multiplication of occasions to give voice appears as an extension of existing arrangements and not as the first stage in their transformation. The aim is the survival at any price of that improbable but irreplaceable being, the ordinary citizen. Provided that from time to time he agrees to become silent again and to accept the rule of delegation, a vociferous, shouting, angry ordinary citizen, organizing leagues and intrigues, is preferable to the contagion of uncertainty that results from the ceaseless calling into question of representatives and the voices they claim to speak for.

What is valid for the ordinary citizen also applies to the layperson. From time to time the latter is worried about what the specialists in white coats are hatching in the silence of their laboratories and research departments. Are the professional researchers and top-flight engineers working for the common good? Are they really sure about the facts and machines they are producing? Initiatives are taken to calm these anxieties whose legitimacy increases the more they seem to be well founded. It is decided that science is a show and open days are organized for laboratories, thus revealing the remorse felt for keeping them closed in ordinary time; the results of research are popularized in order to show that, certainly, researchers research, but they also discover and invent; media events are organized so that no one is unaware of the great contribution that mathematics has made to the progress of humanity and to show that there is still a long way to go before all the mysteries are clarified; laypersons are invited on to administrative councils of research bodies or hospitals; research programs are organized in close consultation with labor unions or users' associations. All of these initiatives make the wound inflicted by the break between specialists and laypersons more bearable, they strive to bring the two sides of the wound together, the better to suture it. But they do so in order to save what seems to be one of the best safeguards against the disorder that could be introduced by the sudden irruption of uncertain knowledge in public.

The Double Exploration of Possible Worlds and of the Collective

All these prostheses which bring their assistance to delegative democracy are good in themselves. They attenuate divisions and make the double delegation livable and bearable. But for all that they do not preclude hybrid forums, no more than they organize them. The overflows to which they give a form are so extensive that they cannot be contained by makeshift remedies. If ordinary citizens and laypersons organize hybrid forums, it is to challenge the double delegation, and with it all the solutions whose only aim is to save it. Even if it is useful, it is futile to make the laboratory partitions more transparent so that the layperson can look through them to see the specialists busy at their work, just as it is futile to offer the ordinary citizen more space in which to express himself. It is these two figures of ordinary citizen and layperson that are in question. With great difficulty, the discussion leaves the spaces in which it was contained. The double muzzling of the double delegation breaks down. A new social space is conquered which will enable new configurations between knowledge and politics to be explored in a way that faces up to the uncertainties weighing on possible worlds and the composition of the collective.

What could happen if we loosened the constraint of debate confined within the restricted spaces of secluded research and representatives designated by ordinary citizens? To answer this question we need only allow the spread of hybrid forums and follow them in their exploration of new territories. As noted in chapter 1, this exploration is undertaken simultaneously in two directions. First, it is an endeavor to identify the problems to solve, and to conceive of possible and acceptable solutions. Second, on an ongoing basis, it draws up an inventory of the groups concerned by these issues and of the identities at stake.

As far as problem solving and knowledge production are concerned, we have seen that, far from leading to a dissolution of laboratory research, the challenge to the break between specialists and laypersons leads to its insertion within a broader continent in which secluded research and research in the wild both find a place and in which rich vascularizations develop through which each is nourished by the contributions of the other. Chapters 2 and 3 have shown that, when cooperation exists, and obviously not without conflict, it may be more or less deep and intense depending on whether there is collaboration at all three of the moments we have distinguished or at only one of them.

The minimal form of cooperation concerns only the return of secluded research to the world, which we have called *translation 3*: adaptation to the complexity and particularities of the contexts of application, and the conditions of implementation of laboratory results, generally require the active contribution of those concerned. Either the white coats, helped by their political allies, get through by sheer effort by taking the risk that this may break down, or else they agree to compromise, to make concessions—that is to say, cooperate with the concerned groups.

The second point of encounter and collaboration corresponds to what we have called the formation and organization of the research collective: ensuring that this collective is armed with all the human and non-human skills that allow enrichment of the knowledge produced, but which equally encourage all the debates and controversies that enable the knowledge produced to acquire its soundness.

The third terrain of cooperation between research in the wild and secluded research is that of the identification, formulation, and negotiation of the problems on which the work of investigation will be brought to bear. (See figure 4.1.)

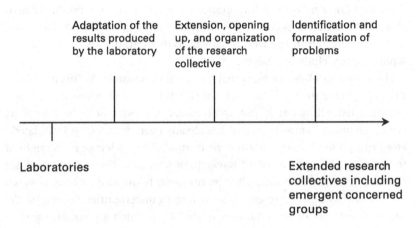

SECLUDED
RESEARCH

COOPERATION
BETWEEN SECLUDED
RESEARCH AND
RESEARCH IN THE WILD

Adaptation of the   Extension, opening   Identification and
results produced   up, and organization   formalization of
by the laboratory   of the research   problems
                           collective

Laboratories   Extended research
                                                       collectives including
                                                       emergent concerned
                                                       groups

Figure 4.1
Different modalities of exploration of possible worlds relative to the degree of collaboration between secluded research and research in the wild.

So there are three different possibilities, three distinct forms of cooperation, each of which can be situated on an axis going from downstream to upstream of the processes of research. Either laypersons are content to wait for the researchers outside the doors of their laboratories in order to convince them to work with them on the adaptation of their knowledge and techniques or they insert the laboratory in a wider collective, introducing new skills and working out a place of their own within it; or they organize the dialogue and exchanges even earlier, even before the researchers close the doors of their laboratory on them. It is a mistake to still speak of laypersons in such configurations: in order to make all the traces of dissymmetry disappear, including and first of all in the vocabulary, it is clearly more correct to speak of secluded researchers and researchers in the wild and to describe three forms of relations between these two populations.

When we move along this axis from left to right, several transformations take place in the regime of the production of knowledge. The first change is in the intensity and depth of the cooperation between secluded research and research in the wild. In successive stages, research cut off from the world of laypersons is replaced by forms of organization that establish an increasingly close association, and at increasingly early stages, between researchers in the wild and secluded researchers. At the same time, we pass from a configuration in which scientific uncertainties are managed by the specialists (whom one asks to come back with certainties), to configurations that grant increasing importance to research in the wild. Movement along this axis corresponds to a change of regime, a qualitative mutation: to its left, research in the wild is denied; to its right, it is recognized in the same way as secluded research.

The reader will have noticed that we speak of research. As this is the custom, we reserve the word 'science' for situations in which research is completed: science is what is not reconsidered (unless to clarify, complete, enrich, or amend knowledge that has already been disputed and validated). From this point of view, whether one is situated to the left or to the right of the axis, one is in the world of research, of science in the making, and not of made science. What does change, however, is the scope of the research activity and its capacity to cope with emergent uncertainties. Delegative democracy gets rid of uncertainties related to research by confining it to secluded laboratories, but in so doing it deprives itself of a powerful tool for investigation: collaborative research, the only one that can fully explore these multidimensional uncertainties.

We can be even more precise. The first step in the direction of an organization of research, which establishes parity between secluded research *and* research in the wild, is evidently recognition of the existence of secluded research. The use of this notion, or of an equivalent notion, actually signals a double recognition: recognition of the crucial role of research, which precedes science, and recognition of the specialized, esoteric, and therefore amendable character of the forms of knowledge that result from it. In talking of secluded research, the recognition that what counts in science are not so much the final certainties as the path followed in order to overcome uncertainties is explicitly acknowledged.

What also changes when we move along this axis is the relative definition of the local and the universal. The vocation, the final objective of secluded research, as the sole mode of organizing research (and its closeness to finished science, which is seen as being intrinsically universal, underlines this), is to produce universal knowledge. As we saw in chapter 2, the paradoxical corollary of the tendency to the universal is hyper-localization in time and space, the extreme seclusion of the conditions of the production of knowledge. Maximum collaboration between secluded research and research in the wild (which thus includes the three modes of cooperation), is, in contrast, entirely aimed at the production of knowledge whose generality is nourished by the consideration of idiosyncrasies and local specificities: the "universal" sheep of the Sellafield experts is replaced by a multitude of sheep, both those of the shepherds near the plant and those raised by other shepherds in other places, a multitude that makes up a richer and more varied, and at the same time more realistic, truthful image of what is designated by the generic term 'sheep'. A sheep that has been studied, dissected, and manipulated by a consortium of both researchers in the wild and secluded researchers would be closer to the sheep of the Sellafield shepherds, since the problems it raises would have been taken into consideration at three points, while also being close to other sheep, since it would have been transposed into and translated within secluded laboratories. There is a reversal of priorities in comparison with simple secluded research: what matters is not the construction of a universal by standardization, and so by the elimination of local specificities, but the construction of a universal through the recognition and successive reorganizations of these specificities. To put it in the language of economic markets (and the analogy is not without foundation): the standardization of mass forms of knowledge gives way to the production of specific, tailor-made knowledge. "You can choose the color of your car," Henry Ford said, "so long as it's black!" "Do what you like with your sheep," say the Sellafield experts, "so long as you follow the general rules we have developed in our research centers." "Choose the model and the provision of services best suited to your particular needs!" says Renault. "Develop rules of conduct, which first and foremost are good for your sheep, on the basis of the knowledge accumulated from other sheep," scientists who have agreed to collaborate with research in the wild would say. This process of investigation, envisaged from the point of view of the successive translations that it performs (chapter 2), can be described as an exploration of possible worlds and of the human and non-human entities comprising them. New identities are proposed, as in the case of muscular dystrophy patients who discover that their fate is connected to the existence of genes which until then were invisible and which they are going to have to take into account.

The exploration of a new space of cooperation between secluded research and research in the wild is made possible by lifting the bans which delegation to specialists in the production of knowledge inflicts on the whole of the social body. We have seen that a second delegation is called into question in the hybrid forums: that which gives birth to that emblematic couple of representative democracies, the strange couple formed by the ordinary citizen and his double, the elected spokesperson who ends up becoming a professional of representation. The second requirement to which the existence and multiplication of hybrid forums testify is the requirement to place uncertainties concerning the composition of the collective at the center of debate instead of relegating them to the enclosures of parliaments and assemblies. How can we describe this movement by which the gap between ordinary citizens and their representatives is not only reduced but reconfigured to the point that the two notions end up losing a part of their pertinence? How can we account for the mechanisms by which the identity of the groups making up the collective and the very composition of the collective are left open for debate?

The approach to be followed in answering this question is no different from the one we followed in deconstructing the separation between specialists and laypersons. Here again, we should reconstruct the different configurations that enable the break produced by this second modality of delegation to be overcome, and indeed effaced. The uncertainties affecting the collective concern the identity of emergent groups, the capacity of each of these groups to perceive the existence of other groups and to take them into account in its own action, and finally the will and possibility of arriving at the negotiated composition of a still unknown collective. We can thus distinguish three stages on the axis that visualizes this work of exploration of the collective and the progressive broadening of the consideration of uncertainties that it generates.

The first stage corresponds to the formation of both specific and supraindividual identities. In fact, one of the most immediate ways of underlining the limits of the delegation by which ordinary citizens leave things to their representatives is to challenge the existence and relevance of this improbable being, the ordinary citizen. The latter gives way to emergent groups, to singular collectives whose identity, composition, and borders are only gradually clarified. In this process of the definition and stabilization of identities, the designation of spokespersons who can be removed at any time and who are in constant interaction with the group is crucial. Identity results in fact from a process of progressive identification that permits the play of mirrors that is set up between representatives and those represented. A group never arrives fully armed as a gift of God. It tests itself, feels its way, and searches for an identity, navigating in the midst of uncertainties. Elements of stability gradually emerged in the subtle dialogue it sets up with its representatives, who can be dismissed at any time. Bit by bit it becomes easier to give unambiguous answers to questions such as: Who makes up the group? What are its projects, expectations, and interests? How does it define or describe itself? When identities are uncertain and still being formed, they can take shape and be constituted only in the constant and changing interaction between representatives and represented. The representative does not record an already existing voice. In fact, the group exists only through the delegation of a voice that it constructs at the same time as it delegates it. Under these conditions, the confrontation between isolated, individualized citizens separated from each other can only be an unacceptable obstacle to the outpouring and unfolding of this maieutics through which unexpected identities are formed. Ordinary citizens, as individual agents constituting a well-defined collective, may exist at the end of this process, but not at the beginning.

The second stage in the process of exploration of the collective allows the formation of identities to go further. It goes beyond the pure and simple assertion of an identity in the process of being formed and suffering from lack of recognition, beyond the single demand of a singularity occupying the whole political space and whose sole obsession is to be seen and heard. The group is no longer content to repeat in every possible way "we are the residents of the future site of the burial of nuclear waste," "we are the parents of children suffering from spinal amyotrophy," and so on. It expresses a willingness to establish dialogue and discussion with other emergent or constituted identities, with other exacerbated singularities, and other groups in the process of formation. In the course of the first stage, silent people recover their voice, not so as to renew an individual dialogue with their representatives, but in order to launch themselves into a collective dynamic with initially barely defined contours, but which, through successive sequences, may lead to the clarification of who they are. These mutes who have become talkative again, both amongst themselves and with their spokespersons, express themselves, but, inasmuch as that is all they do, then they remain stubbornly deaf: what matters is that they are heard and not that they listen. The second stage is when they regain their sense of hearing. Anglers, farmers, the tourism industry, local communities, angry residents, and heritage associations expressed themselves on a redevelopment project for the river Arc. Although some of these groups were already formed before the project was announced, others are its product so to speak. Boisterous groups emerge which change their views from time to time, put forward arguments that are difficult to follow, and suddenly disown a representative who was previously considered legitimate. And then, instead of being content with proclaiming their identity, enclosed in their own universe, marked by a sort of political autism, they begin to argue amongst themselves and recognize the multiplicity of groups, be they emergent and unexpected or entrenched and well known. How is this opening brought about? We will answer this question later. It is sufficient to note here that unstable groups, focused on themselves, are succeeded by groups that are just as emergent, but which are prepared to listen to other groups and perceive their discourse, prepared, in short, to recognize their existence and identity. As a result, they discover that they share their history with others and that, in struggling to get their own (evolving) identity recognized, they are also struggling for recognition of the identities of other nascent (or even established) groups: if they were not to listen it would prove in advance that those who refuse to listen to them are right.

The third stage then opens up, in which the clash of singularities, expressed and listened to, gradually leads to their composition. How can a necessarily provisional collective be formed that takes each group into account, whether it is an emergent group or one already formed? To find an answer to this political question par excellence, we need to go beyond just the expression of views, and we need to go beyond just listening, however attentive and empathetic it may be. Each group, either emergent or established, must accept that its own identity is negotiable, and that the composition[8] of the collective requires compromises and adjustments with the other identities involved.

Asserting one's emergent identity by expressing it strongly and clearly, listening to and recognizing other emergent or already existing identities, and then entering into their discussion and intersecting composition: these are the three stages that mark out a route which diverges from the *mise en scène* of the ordinary citizen and his or her spokespersons. The collective is composed (and no one knows how it is composed until the end of the procedure), instead of being seen as no more than the result of an aggregation of individual wills (a result which varies according to the procedure of aggregation employed). The very idea that there are particular individuals on one side and, on the other, a general will manufactured by means of statistical instruments, is seriously shaken with the emergence of groups that assert and define their own identity and then, when they discover that they are not alone, agree to debate the composition of the collective in common. In this shift, the idea of an ordinary citizen who possesses inalienable individual rights is not lost on the way. The same is true for relations between the aggregation of the collective and the composition of the collective as for those between secluded research and research in the wild. The emergence of groups, their mutual recognition, and then the composition of a collective allowing each to find a recognized place, no more obliterates individual rights and the construction by aggregation of the general will

than research in the wild aims to obliterate secluded research or take its
place. In both cases, what is involved is enrichment, going beyond the
mechanism of delegation to the advantage of a more symmetrical and bal-
anced involvement. Left to itself, research in the wild would be cut off from
the extraordinary power of translation and amplification that can be provided only by secluded research. In the same way, if the composed collective were not reshaped according to the procedures of the constitution of
the aggregate collective, it would not be able to produce the individual citizen, on one hand, and a general will, which is not just the will of the
strongest, on the other. Conversely, cut off from the potential for transformation and reconfiguration made possible by the open and unsettled
composition of the collective, procedures of aggregation, left to themselves,
close the door on the recognition and consideration of emergent singular
identities.

These considerations clarify the status of the second axis that will serve
to delimit the space in which hybrid forums take place (figure 4.2). The

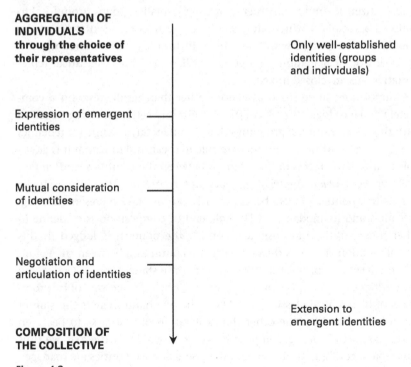

**AGGREGATION OF
INDIVIDUALS
through the choice of
their representatives**

Only well-established
identities (groups
and individuals)

Expression of emergent
identities

Mutual consideration
of identities

Negotiation and
articulation of identities

Extension to
emergent identities

**COMPOSITION OF
THE COLLECTIVE**

Figure 4.2
Different modalities for defining the collective relative to the degree to which emergent identities are taken into consideration.

intensity and depth of the movement of the collective's composition increase as we move down this axis. The empty space which was extending between the ordinary citizen and the aggregate collective is now being inhabited by emergent groups. Initially they are solely concerned with their existence, but then become increasingly active in discussions and negotiations with other groups, be they emergent or established, and in defining the collective.

The reader will have noticed that we have used the notions of *aggregation* and *composition* to describe the contrast between these two regimes of formation of the collective. These two notions actually express the difference (and complementarities) between the two regimes. Aggregation, in fact, presupposes the existence of indisputable basic units that are identified with individuals, the ultimate and fundamental entities starting from which and on which the collective is constructed. It also presupposes a process of classification, grouping, and hierarchical organization, which is possible only on the basis of these indisputable entities. This relies on statistical techniques that, starting from a large number of distinct units (individual citizens), aim to construct increasingly smaller groups (from local to national assemblies) which are nonetheless seen as representatives of the population one started with and which, in turn, can be summed up, still by means of a calculation, in a general will: political and statistical representativeness are closed linked.

Composition, understood as action rather than result, rests on a completely different logic. It replaces the classificatory certainties of aggregation with the uncertainties of groupings that simultaneously define (or redefine) the significant entities, those that are able to speak and to whom it is advisable to listen, the forms of the relations between these entities, and, *in fine*, their *modus vivendi*. Aggregation does not reconsider the entities to be aggregated: political debate bears precisely on the stake represented by aggregation and its modalities.[9] The sole end of composition is to define in what these entities or substances consist: the political is lodged in this reconfiguration. The axis thus establishes a continuum between the voluntary management of uncertainties regarding the state of the collective and the entities of which it is composed, and their being taken care of by procedures of hierarchical classification.[10] On the one hand there is the sum of the collective and on the other the exploration of who asks to be taken into account in order to compose it. As we move down the axis we thus pass from a configuration in which the political uncertainties are managed by a handful of chosen (elected) representatives who have all the time to debate, but who, in accordance with mechanical procedures, will end up imposing a will that will become *ipso facto* general, to a configuration in which these uncertainties are dealt with by multiple emergent groups.

When we move along this second axis, not only does the distinction between the ordinary citizen and his representative disappear, but also the modalities of the distribution of singularities and the relations established between them are transformed. In the pure regime of collective aggregation, each individual endowed with his own preferences, interests, or will is supposedly irreducible to every other individual. He is a will that cannot be absorbed by anything other than that which it has laid down. As we have seen (and on this point Rousseau's intuition turns out to have been verified), the possibility of a common will arises precisely from the extreme diversity of the citizens.[11] But, as citizen, he is similar to every other citizen, since he possesses the same rights and is endowed with the faculty of choosing what he wants and of wanting it in complete autonomy. It is this formal equivalence between each individual, between each citizen, that enables us to say that each counts as much as any other whomsoever, that each individual's voice, singular certainly, is only one voice among others with the same weight and deserving the same consideration. This is the basis on which the general will emerges: since each voice counts in the same way, it suffices to count them, by grouping them in terms of their singularities, so as to bring to light what counts for the collective considered in its totality. The general will, which is valid for all and for each, and which is the equivalent of the property of universality for knowledge, is produced on the basis of singularities and individual specificities on which, once it is formed, it "falls back," producing uniformity where once there was the most extreme diversity. The election, the expectations, and the infinitely complicated calculations to which it gives rise, is thus a formidably effective set-up for aggregating and reducing a great number of wills, each different from the other, each possessing the right to participate in the definition of the general will, as a single will that is no longer tied to any particular individual.

In the regime of the composition of the collective, singularities are asserted and claimed, instead of being erased, and the affirmation of their content constitutes the very substance of political debate. The contrast with the regime of aggregation of the collective is striking, since the latter works desperately to obtain the bracketing off of singularities while relying on them for defining the general will. First, the latter are reduced when the microcosm constituted by representatives replaces the macrocosm of the population of ordinary citizens. Then they are reduced again when these representatives come together to debate and form a general will (the law) that constitutes the collective as sovereign and refers particular wills to contingencies with no political importance. What is lost en route is the flesh and blood of particular identities. Speech becomes increasingly political in proportion to it being purged, by fractional distillation we could say, of its individual problems and local considerations. Procedures for aggregating the collective make possible the expression of a general will, for they get rid of particular voices and of what makes each of them the authentic voice of a housewife in her fifties, of a retired colonel's wife, of a worker on a Ford's assembly line, or of a secondary-school teacher from Eau Claire, Wisconsin. Instead of making every effort, with great persistence, to remove these singular identities in the aggregation of the collective, one strives to conserve them, preserve them, and restore them, with equal persistence, in the regime of the composition of the collective. In the latter regime, instead of counting votes that have been rendered formally identical in order to reveal what are described as the more profound resemblances behind secondary differences, what matters in fact is being interested in what is specific and singular in particular voices in order then to compose them without concealing their existence. A universal (the aggregate collective) obtained through finicky elimination of specificities is replaced by a universal (the composed collective) linking singularities that have been rendered visible and audible.

One cannot fail to be struck by the homology of the transformations that take place along the two axes being considered. In both cases, what is called into question is the production of two populations and the breach between them. Here we witness the appearance of groups of patients who mean to take an active part in the production of knowledge at the same time as they assert their wounded identity; we discover angry residents who speak of their difficulties and fears and who designate spokespersons to take part in technical discussions. Together they mark out the existence of a new territory, a new political stage, which can be described and mapped out with the help of the two axes presented above. This space, which reveals to us the hybrid forums and their overflows, communicates with the old scene of secluded research and of the aggregated collective.

Figure 4.3 clearly illustrates what distinguishes delegative democracy from dialogic democracy. The former is held in the upper left quadrant whereas the latter extends down into the lower right part of the diagram. We pass from one to another gradually, combining the different modalities of the exploration of possible worlds and of the constitution of the collec-

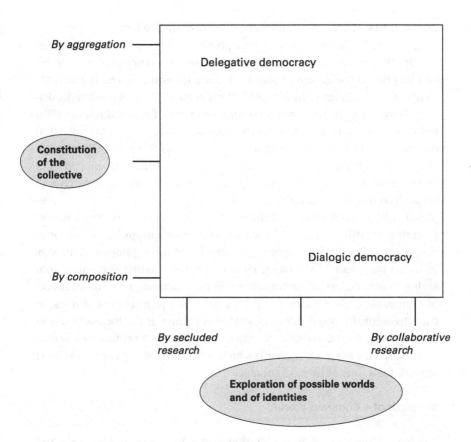

Figure 4.3
The dialogical space.

tive. It will be understood that in its excessive simplicity this diagram does not exhaust everything that could be said about the democratization of democracy. It confines itself to noting, on the complicated map of the procedures of which democracy consists, the new lands conquered by hybrid forums. Merely by their existence, the latter show the extent to which the double delegation is an obstacle to the political treatment of uncertainties; the overflows they set up cannot be contained and colonize previously unexplored social spaces. They install themselves on the terrain of dialogic democracy; it is there that they devise procedures and forms of organization that will interest us in the next chapter. The more actors venture into this space, the more they distance themselves from the upper quadrant of the diagram, and the more they are able to cope with deep and productive uncertainties. There are uncertainties concerning scientific knowledge, which make it difficult to describe the possible worlds or scenarios between which choices have to be made, and there are uncertainties about the list and identity of the groups in search of existence and so about the possible forms of the collective. In the gradual movement from delegative to dialogic democracy, the cold but reassuring certainty of finished science ("Do not come back to us until you have certain knowledge") together with that of a general will formed by successive aggregations ("Vote, we will take care of everything") gives way before the exciting but disturbing uncertainties of an investigation involving cooperation between secluded research and research in the wild that cannot be entirely programmed, and a never completed work of composition of the collective on the basis of continuously emerging identities. Accepting the participation of groups in the composition of the collective, and agreeing that the list of these groups and the way in which they define their identities may fluctuate, means abstaining from saying in advance what the collective will be. Tolerating the multiplication of sources of problematization, the extension and restructuring of research collectives, or the proliferation of strategies aiming at the local adaptation of knowledge with a vocation to universality, means accepting in advance that the worlds in which and with which collectives will be composed must remain, for a time at least, negotiable.

In Search of a Common World

The two dimensions of an exploration of possible states of the world and an exploration of the collective are closely linked in hybrid forums. But before dealing with the interdependencies created between them, we should show in what respects they should not be confused.

Movement along the axis of research is possible without the modes of composition of the collective changing. Let us return to the inhabitants of Woburn. They do not hesitate to involve themselves fully in detailed cooperation between secluded research and research in the wild, organizing close and early cooperation between the two. This boldness does not lead them to attempt the same venture on the side of the exploration and composition of the collective. On this second front, they are satisfied with asserting, in an unproblematic way, their identity as a group: the inhabitants of Woburn living close to waste the toxic nature of which is responsible for an increase in infantile leukemia. There is no question of them giving way on the definition of this identity or accepting that it can be absorbed in any party political program. They want their singular voice and views as parents wounded in their flesh and blood to be taken into account. It is their children, this one here, that one there, who are dying, and not just any anonymous child living alongside just any toxic waste. In defending their singularity as parents united by the same misfortune, they have abandoned the fiction of the ordinary citizen detached from any group identity. By demanding that interest be taken in their own children they do not transform themselves into a simple pressure group that would strive only to defend its positions without departing from the regime of delegative democracy. Their protest is not comparable to that of pigeon hunters who noisily mobilize against European directives or those of arms manufacturers who struggle for national preference. Certainly, to start with they are struggling for their interests, but if they do so it is in order to be clear about the nature of their interests. When one sets off bag and baggage and enters territories where knowledge is uncertain—and in this case there is extreme uncertainty—identities become emergent. The residents of Woburn face the question of whether their concern is legitimate, of whether or not they should struggle in the name of their children's survival. They have their doubts, but they are not sure of anything, and that is precisely why they embark on collaborative research. At the end of this research they may know whether they are parents whose children are dying from the fact of toxicity or just residents like any others: their identity has been rendered dependent on the course of the investigations. Pigeon hunters or manufacturers of air-to-air missiles are not in this situation; they know that they are struggling so as to be able to take their rifles down as soon as possible from the racks where they have been all summer; they know that they have to convince the political decision makers that it is better for them to buy French. The parents of Woburn know nothing like this, but they know what they want to know. What defines them as a group is this desire to construct an identity that is finally stabilized and which for the moment is fluid and undecided. There is no question of them either returning to delegative democracy or going very far in the composition of the collective. They are completely wrapped up in their struggle for the recognition of their emergent identity. That is why they do not hesitate to go very far on the horizontal axis—as far as is necessary to reduce the anxiety gripping them. For them, the composition of the collective is not yet a primary concern. They have simply understood that delegative democracy cannot take care of them. They have too little visibility, too little audibility, there are too few of them, and their influence is too weak for professional politicians to show any interest in them. But they are still too steeped in uncertainties to take an interest in other groups and to imagine being part of a recomposed collective. While they refuse to budge an inch on the vertical axis, they are not afraid to venture far in the exploration of the horizontal one. Later, when their matters of concern have been identified more clearly, when the investigations undertaken have revealed the formerly invisible entities from which their misfortunes probably stem, when the causal chains have become apparent, then, with new room to maneuver, they will be able to engage actively in the negotiation of the collective.

Symmetrically, it is conceivable that the composition of the collective may advance without any change in the modes of organization of cooperative research. Political philosophy has clearly perceived this when it deals with the classical question of minority rights. In this case there is no need to be concerned about active and deliberate scientific investigations. There are in fact many emergent identities that are only distantly linked to organized scientific exploration of possible states of the world. The cases most of these philosophical works consider are actually concerned with the problem of the place and legitimacy of ethnic, religious, and even quite simply linguistic claims. The case of the identity of the French-speaking community of Quebec, brilliantly analyzed by Charles Taylor, provides a striking illustration of this possibility.[12] Nothing in the constitution of the identity of this group can be remotely associated to the questioning or uncertainties that could be explored by recourse to secluded research, research in the wild, or a possible combination of the two. This does not prevent the actors from pushing reflection on the modes of composition of the collective to the limit. The Quebec case, like that of the chador in some European countries, has the huge merit of posing in its full extent the question of the foundational or non-foundational character of the great division between ordinary citizens and their representatives: should the rights of the Quebec minority take precedence over those of any other French- or English-speaking citizen, and if so, in the name of what principle? It will have been understood that this question is directly linked to that of positions occupied on the vertical axis of our diagram.

The form taken by the collective, as well as its constitution, depends on the answer given to this question. We know Charles Taylor's solution, which Michael Walzer's reflections have extended[13]: the democracy of human rights, that which *in fine* affirms the existence and irreducible preeminence of the ordinary citizen, is included in the democracy of emergent minorities. This is not the place to discuss the pertinence of this answer. (See chapter 7.) What we want to emphasize here is the autonomy of the question of the composition of the collective in relation to that of the exploration of possible worlds, or the independence of the vertical axis in relation to the horizontal axis.

This independence explains the plurality of the possible configurations of hybrid forums or dialogic democracy, since it opens up a space of combination that allows for a great variety of forms of organization and trajectories of development. Before describing this diversity, there remains a problem to be resolved. It is one thing to demonstrate the independence of the two axes, but it is quite another to account for their dynamic interrelation. Nothing is static in a hybrid forum. The identity of the groups that take part in the composition of the collective varies as the controversy develops; forms of organization of research develop in terms of the results obtained. Obviously, the two explorations, of the composition of the collective and of possible states of the world, become entangled. As we will show with two examples, if these interactions are possible it is because there are many cases where, in order to pursue the exploration on one of the axes, actors change level and reopen the discussion by moving on to the other axis. To put it in a colorful way: one way for protagonists to unblock research that has reached an impasse and is failing to produce acceptable results is to abandon the exploration of possible worlds for a moment and to start the discussion of emergent identities and their adjustments. And vice versa. When the work of the composition of the collective has come to a halt, the solution often passes by way of a revival of research within the framework of a closer and deeper association between secluded research and research in the wild. When identities seem to be incompatible, a way of acting on them and of revealing opportunities for compromise is to revive the scientific investigation. When, on the other hand, research proves to be sterile and to hold little promise, a way of escaping the dead end is to renegotiate the matters of concern and their hierarchy, with other concerned groups.

Consider an association that brings together persons affected by serious neuromuscular diseases. Not only are their diseases of no interest to anyone, not even to a doctor or a researcher, and even less to a politician, but their existence is also denied by the multiple exclusions they suffer. It is in order to escape from this state of non-existence that the sufferers group together and create an association so that they are taken into account and not just ignored. They claim the right to expression: "We exist!" is their first intervention in the public space. Delegative democracy, with its double delegation, is powerless in the face of this type of demand. The obvious strategy for a group which is still weak, barely visible, and what is more unable to mobilize broad and powerful social networks quickly, is to link its struggle to that of well-known action groups which already have the ear of public actors. In order to be heard, gain numerical strength, and not disappear in general indifference, why not underline similarities with other handicapped groups and make common cause with those suffering from motor disabilities, the victims of road accidents, or even older persons who find it difficult to get around? Is not the most important thing to slip into existing movements and merge with constituted groups and identities so as to avoid exclusion? If they agree to deny their identity as victims of myopathy and to redefine themselves as suffering from motor disabilities craving integration, they would slip into existing categories while taking care not to challenge the balances of delegative democracy. This work of de-singularization ("We belong to the already constituted big family of those with motor disabilities") seems particularly easy and more satisfactory when the group's specificity ultimately hangs on so little: after all, it is tempting and defensible to identify someone affected by limb-girdle muscular dystrophy (that rag-bag category of myopathies) with someone affected by multiple sclerosis or poliomyelitis. Are they not brought together by their inability to perform certain actions and their shared demand for technical or human assistance to enable them to live like everybody else? However, even if this solution of integration within preexisting categories is tempting, it ends up leading the group to self-censorship and to severing a part of itself. The needs expressed by a child who has undergone a tracheotomy (and who can express himself only through a voice synthesizer or by manipulating the joystick of a computer which acts as his apparatus of phonation) are not in fact exactly the same as the needs of adolescents placed in the category of motor and cerebral disabilities who, being unable to articulate intelligible phrases and condemned to express themselves in mumbles that only those who are trained can understand, must be constantly accompanied by their personal "translators." The association very quickly understands that in order not to be lost in populations that are too large and too different, and which, because they are already constituted as pressure groups, cannot be modified, it must make its voice heard and make its specificities audible. The association realizes that it has to explore new worlds in which those suffering from myopathy would have an unquestionable, quasi-objective identity that they would be able to express, explain, and articulate in a public space in which they would be listened to and recognized.

The challenge thrown down by the leaders of the association is to objectify what are felt to be singular subjectivities, but which cannot be expressed or conveyed before being objectified. There are not a thousand ways in which this challenge can be taken up. How can we know if a world could exist in which those affected by neuromuscular diseases would be no more than those affected by neuromuscular diseases? The uncertainty is complete. To remove uncertainty and bring about this common world in which those suffering from myopathy would find their place in a collective they have composed with others, the only solution is to embark on the exploration of possible worlds, to work at the coalface of investigation where, after having identified and formulated the problems (what is the cause of myopathy?), you organize research in a way that will provide elements of an answer. There is no point in waiting with arms folded outside the laboratory, because there is no laboratory working on the subject; there is no point in struggling to extend and structure a research collective that is conspicuous by its non-existence. There is no other solution than to begin at the beginning, with what we have called "the primitive accumulation of knowledge." Observations are collected and knowledge formed which will provide a solid basis to support and be developed by laboratory research. Patients and their families do not stop there; they accompany the researchers in their laboratories, forming DNA banks and then cell banks, setting up structures which enable collective experimentation to be organized, followed by drug trials. The exploration of possible worlds advances in strides, co-piloted by laboratory researchers and those effective and fearfully pugnacious researchers in the wild, the patients. In the process, those suffering from myopathy are able to construct a new identity that cannot be reduced to or absorbed into any other identity, an identity as objective and real as the genes and proteins that are its cause. The young children suffering from a spinal amyotrophy that ends up destroying the motor neurons and cuts off all communication between the spinal cord and the muscles it controls, leading to, among other misfortunes, a progressive asphyxia, are no longer human beings like others. They belong to the group of those whose gene SMNr, situated on the long branch of chromosome 5, has suffered serious deletions. They are clearly human beings in their own right—who could now doubt it?—since to a few base pairs their genome is almost identical to any other human being and the existence of this muddle is the simple result of natural random processes which have nothing to do with any human will or project. They are nevertheless particular, singular human beings, since one of their genes is modified and the result of this is underproduction of a protein without which the motor neuron cannot survive. The exploration of the genome, to which the patients have made an irreplaceable contribution, reveals this double objective fact:

that of belonging to common humanity, and that of the existence of a genetic feature that brings out a specificity in an unchallengeable way. Those suffering from SMA draw their irreducible singularity from this: their identity, the feeling of belonging to a different group, but one which is attached to a wider collective, is inscribed in sequences of bases that can be read on a computer printout.

This identity is at once subjective and objective. It is objective since, *in fine*, it is a question of biochemistry. No one could deny this self-evidence: a child with SMA is such because his SMN gene is incomplete. Identity is therefore not reducible to an imaginary representation; it cannot ignore this objective reality. But this does not take away its subjective dimension: this identity is not imposed on those suffering from myopathy from the outside. They develop it at the same time as they discover its objective component. It is in the simultaneous management of these two dimensions that cooperative research shows all its effectiveness. At the same time as it is established on the biological level, genetic specificity is quickly translated into social identity, feeding a process of differentiation. All the ingredients are present for the production of this mutation. The patients are joined together in tracking down the gene, in localizing it and identifying it. By taking an active part in this investigation they are launched into what we should call an enterprise of introspection which is certainly equipped with tools, but which has the sole aim of showing, both for oneself and others, what one is. Γνῶθι σεαυτόν (Know yourself)! Those suffering from myopathy are zealous adepts of the maxim inscribed on the wall of the temple at Delphi. But they know that there is no introspection that is not equipped with instruments of one kind or another. Intellectual exertion alone is not enough. Even in its most native, most immediate forms, introspection calls upon bodily techniques of concentration upon oneself. And when it sets its sights on invisible genes it cannot do without with heavy investments. On the axis of the exploration of possible worlds, the meaning and aim of the movement of those suffering from myopathy is to create the infrastructure necessary for tooled up introspection. What they reap at the end of this anxious and costly quest is a new definition of themselves irreducible to any other: because they are the carriers of an injured SMN gene, they "are" SMA, people with spinal muscular atrophy, both human beings similar to every other human being, yet profoundly different from the great majority of them. The work undertaken on the axis of collaborative research has led them to move along a notch on the axis of identities and their composition. Rather than melting into the crowd of disabled people, they are in a position to express their singularity, that of an emergent identity, without for all that dissociating themselves from the disabled.

After this detour through cooperative research, those suffering from myopathy find themselves endowed with a specific identity. They are ready to embark perhaps on the adventure of the intersecting discovery of identities: "We are SMA, with our specific features, but since we are humans in our own right, we cannot simultaneously assert our humanity, demand that our singularities be recognized, and not recognize the singularities of all minorities who assert themselves as both different and similar." They are present on every front explaining their new identity, describing themselves, and not allowing anyone else to describe them. They do not seek to occupy the entire stage in this work of self-presentation. They know that their identity will be more firmly established if they make room for other identities. This tolerance is not a sign of weakness; it is awareness of one's own strength. It is the simple consequence, the simple profit from a costly and sophisticated introspection. How many genetic mutations and muddles are possible? As many as there are genes, as many as there are proteins expressed, and even more, for disorders linked to a single gene are the exception. In the almost infinite combinatory of genetic accidents, SMA is only one possibility among many. Without leaving the human condition, and content with reading the genome, the investigation undertaken by sufferers from myopathy not only provides each neuromuscular disease with its specific identity, its personal signature, but in addition it establishes this identity as one among many different as well as comparable possibilities. In this way, the involvement of myopathy in the enterprise of exploring possible worlds brings with it the recognition of other, equally improbable and unexpected identities and, at the same time, a potentially infinite tolerance toward them.

We began with the problem raised by the rigidity of the double delegation and its inability to allow for the expression of inchoate, emergent, and evolving voices. Within delegative democracy these voices find no representatives prepared to listen to them, to take them into consideration and to serve as legitimate spokespersons on their behalf. As we have seen, delegative democracy implies that arguments, interests and expectations be established with sufficient stability. By bridging the gap between ordinary citizens and their representatives, it can give a place and a voice to groups formed around well-identified and well-established causes. These action groups, also known as pressure groups, often powerful and well organized (they may be firms, trade unions, NGOs, or religious organizations), are prepared to fight to defend their interests and positions, which emergent groups are likely to threaten. They are "already there" and have often been established for a long time. The spokespersons of these instituted groups have no difficulty silencing their troops and speaking in their name, because the lines to be taken are so predictable. The constituted action group is known and recognized: unlike the emergent group, it does not have to struggle to get its voice heard and listened to. It is already in the political field. Moreover, it only pursues the logic of aggregation, accentuating it. It constitutes a first-order aggregate, bringing together individuals who are supposed to be identical and who keep quiet so that their representative speaks in their name and speaks more clearly and forcefully because they are many. With the constituted action group, we are already, we are still, in the realm of the politics of large numbers, since the group's ability to exert pressure is directly linked to the weight of its membership. But, as a group, it nonetheless underlines the limits of the pure model of aggregation for which the only basic element is the ordinary citizen: it is the outcome of a sort of secondary aggregation weighing on primary elective mechanisms. In the purified model, intrigues and cliques are suspect, for one never knows whether they are composed of free or subjugated individuals. Accepting the well-established action group means tolerating an infringement of the principle that only individual citizens exist. This is an especially serious infringement in that cliques or leagues are rarely models of internal democracy! By allowing the constituted group to assert its interests and proclaim them loudly and clearly in a public space in which it claims its place, and allowing this without any control over how its members' consent has been obtained, delegative democracy opens the door to new forms of collective co-positioning—only to shut it again immediately. By transgressing the sacrosanct rule of individualism, delegative democracy actually prepares the way for the logic of composition, which becomes more inevitable when individual wills are still inchoate and fragile, withdrawing in favor of the collective identities on which they depend.

This strategy of adaptation of delegative democracy pertaining to the construction of the collective is symmetrical to that which it developed with regard to scientific investigation. When they are powerful and influential, established groups have a say, either to weigh on the orientations or organization of research or to ensure that the common world which they prefer prevails. This "opening" simply shifts the gap: the two great divides underlying delegative democracy are maintained. But a breach is opened. Once the groups are admitted, once the doctrine of double delegation is transgressed, it is difficult to stop midway. Why this group and not

another one? From what point is a group no longer emergent? Why penal-
ize those that have not yet stabilized their resources and identities?

The ongoing adventure of French myopathy patients is a perfect illustration of the mechanisms behind the subversion of these boundaries and the
construction of the sphere of dialogic democracy. By refusing, notwith-
standing pressure, to join an influential and powerful established group
(i.e., disabled persons) which has considerable means within delegative democracy for weighing on the construction of the collective (for instance by
fighting for full accessibility) or for orienting technological innovations,
myopathy patients show that they are exploring other forms of democracy,
those which recognize that emergent identities have the right to simultaneously engage in research and shape the collective. We can easily imagine
the difficulty of this extension of the democratic territory: the institutions
of delegative democracy and the established groups that have learned to
take advantage of it cannot be favorable to such subversion of the great
divides that are the mainstays of their power and stability. If the AFM is
able to escape from the ascendancy of delegative democracy and its established groups, it is because it deliberately and boldly embarks on the dual
exploration of possible worlds and conceivable collectives.

Before the investigation, the inherent logic of delegative democracy
tended to impose a double reduction on them: first, the reduction of myopathy to a handicap, and second, the reduction of the handicap to its care
which aims to integrate the handicapped person and allow him a life
judged to be normal. After the investigation, a turnaround has been accomplished: those with myopathy could not be reduced to any other definition
than the one they give themselves. This sudden turn brings about a radical
reversal. The change of regime justified by the double exploration is that of
adapting the collective to those with myopathy rather than the other way
round. Thus, through successive repetitions and movements back and
forth, a common world is formed made up of mutated genes or genes
riddled with deletions, of identities constructed on the basis of missing
genes, and then around groups, all with disabilities, but each suffering
from specific and different deficiencies. No one could anticipate this common world, not even the concerned and emergent groups. It is the fruit of
an entangled quest. And in addition at the end of this difficult introspection, not only are those with myopathy in a position to impose their voice
without it being distorted, but they are also working for the composition of
a collective that is more welcoming to the great variety of disabilities. Thus,
with dual exploration new identities can be added to the existing list, and

their compatibility guaranteed. Production of new identities, articulation and composition of identities: it is clear why we talk of a common world. Dialogic democracy is the name given to this dynamic process of constitution of the common world, which is a deliberately open, future world.

It is clear that the production of such a world, both constructed and quite real, would have been impossible without the development of new forms of knowledge that bring about the emergence of non-human entities like the gene. In joining in the history of the patients and their diseases, these entities open up the field of the possible. Genes are not external to the exploration and composition of the collective. They are directly involved right from the start. At no time do they present themselves at some border to ask, through the scientists, if they can enter into the composition of the collective. Naked, isolated, and unattached genes do not exist. From the moment they enter the picture they are escorted by the patients associated with the researchers. They are caught up, enrolled in the production of the identity of those with myopathy. They emerge into existence at the same time as they make their contribution to the emergence of new groups. Contrary to what a too hastily naturalistic vision might think, the gene operates on two axes: that of the research into possible worlds and that of the composition of the collective. It is through its intervention that the initial problem—constructing a common world in which someone with myopathy can be someone with myopathy—finds its solution. Discovering this world, or rather producing it, required this work of exploration made possible by the organization of collaborative research.

Those suffering from myopathy help us to understand the interest of the notion of a common world. Through their action and the exchanges in which they take part, they lead to the existence of a new world which is profoundly different from the one in which they previously had to survive. We find genes here whose existence no one suspected; we find research laboratories, cell banks, genetic consultancies, care institutions; we find prostheses that compensate for the deficiencies of the patients in their everyday life; we find legal arrangements guaranteeing minimum rights to so-called handicapped persons. A whole world has been explored and constructed step by step, starting from the one that existed in order to enlarge it, transform it, and enrich it by introducing new elements. This world, one among many that could have come into being, has the property of having been negotiated, discussed, and tested in such a way as to transform identities to the point of making them, at least for a moment, compatible with each other. That is why this world can be described as common. It belongs to those with myopathy, but equally to all those who were involved with them in its exploration and development. We can see that such a result would have been out of reach without the dynamic of the double exploration: one brings about the emergence of possible worlds and identities, the other composes them in such a way that each can find his or her place in it, and if no compromise can be found, the return to research may bring to light new options and result in new proposals. No doubt the most general definition of what we understand by a common world is a world with the double characteristic of being quite real, since it is the result of a long process of objectification, and of being inhabited by subjectivities that are adapted to each other and directly involved in this world. We note that at no point was this or that common world wished for as such. The common world is not the consequence of a project which we would find really difficult to explain where it comes from. To account for this world's construction, we need only think of these patients and their families engaged in the quest for their possible identity and, in order to arrive at it, to embark on collaborative research and, one thing leading to another, design other modalities of collective life.

The roads leading to the common world are as many and as confused as the ways of providence. Let us now follow another trajectory symmetrical to the one we have just been analyzing. This time the point of departure is situated on the horizontal axis. Engineers and researchers have given a unanimous answer to the question of what to do with radioactive waste: "Bury it, and bury it as deep as possible!" And they have added "in the Gard, for example, where the geology seems favorable." As soon as this statement comes out in the public space, the reactions multiply. The viticulturists of the Hérault, as anxious neighbors, are against the image of the atom being associated in any way with the wine they produce and export. Even if the site is absolutely secure, they claim, one cannot prevent Japanese consumers from making a link between the locality of the vineyards and the presence of nuclear containers. And if these suspicious Japanese don't happen to think about it themselves, sympathetic competitors won't refrain from bringing it to their attention! In the name of the danger of job losses and the need to maintain the region's economic dynamism, they ask for the project to be reconsidered. The choice put forward by ANDRA brings into existence this very real group—"the-viticulturists-with-commercial-interests-in-Japan"—which was previously formless and now raises it voice to defend its existence and identity by hastening to link them skillfully to the fate of the whole region. The possible world devised by the secluded researchers of ANDRA is violently challenged as soon as they put their noses outside their laboratories. "I wish they'd return to them!" angry viticulturists, ecologists anxious about the environment, and local councilors careful not to lose their electors, cry in unison. "Why don't they integrate into their projects the consequences of their programs on the economic life of the region? Stop the excessive overflows! We don't want this world!" they continue with one voice.

Should one force it through? Repeat, with a hint of annoyance and in the tone of a nineteenth-century primary-school teacher, that something really must be done with this waste? Assert, while vaunting the general interest, that the proposed solution is the one that objectively entails least risk? Continue repeating to whoever will listen that the Gard site offers all the guarantees one could wish for? Intone that the residents must accept the waste in the name of the nation's future and independence? Admit, *sotto voce*, that residents could possibly be given compensation calculated on the basis of a cost-benefit analysis that some consultant economists will hasten to undertake *pronto subito*? Bring out a Nobel Prize winner to change the units of measurement of radioactivity in order to get the somewhat obtuse viticulturists to understand that their vines are subject to much more sizeable natural radiation?[14] Such a clear-cut decision would be conceivable in the case of delegative democracy. It would suffice to vote a law based on the calculations of experts. But what the viticulturists' voice shows is that that framework is no longer legitimate, that the very fact of the researcher's proposal has produced an overflow that has burst it open. Their proposal brought about the emergence of groups which were invisible and unthinkable. The Gard viticulturists had no voice in the matter when the general will decided in favor of an all-nuclear future, and for a good reason! They were not yet living in this world, that is to say, in the world in which one is looking to bury nuclear waste that one does not know what to do with and in which the Gard looks like an ideal repository. For this world did not yet exist as a possible world; not being conceivable, it was not even debatable. That is why it is not easy to disqualify the voice of these viticulturists: they are opposed, indirectly, to a decision that concerns them directly but with which they were not associated! How can a debate be refused? There is nothing scandalous in being indignant about being the only victims of a choice which they were not consulted about! And let's not get our hackles up about the word 'victim'! At no time is it a question of radioactivity, of contaminated groundwater, or of irrational fears in what they say! It is a matter of commercial risks and, through these risks, of the existence of a group, of attachment to one's profession, to one's *terroir*, in short to one's identity. Imagining that these anxious viticulturists will be

reassured by the learned calculations of physicists, geologists, and economists is like pretending to believe that the French miners went on strike in
the mid 1960s because they feared firedamp explosions or the collapse of
tunnels in which some of them lost their lives. What drove them to take
to the streets, in demonstrations that paralyzed the whole of France and
forced the Gaullist power to send in its state security police force, was the
threat of pit closures and the intolerable idea that their existence could be
put at risk. One has to be rather obtuse to think that people engage in the
calculation of risk before making or accepting a decision! If that was the
case, the motorways should be deserted on Easter Mondays! Being a motor-
ist is not being a calculator. As Sartre expressed it with regard to smokers—
to smoke is to exist—we could say that drivers exist as motorists, and that it
is pointless asking them *voluntarily* to cease taking risks by driving. They
would have to repudiate their identity as motorists! It is as if we were to
suggest that human beings put an end to their days rather than wait,
immersed in the uncertainty of the moments flying past, for an ineluctable
end! Forgive our bluntness, but it is only engineers, economists, and insurers who think, first that we decide to exist, and second that this decision
hangs on an explicit or implicit calculation of risk! Our engineers of nuclear energy moreover are only beginning to understand it, and evidently
they have no answers to give to the viticulturists. The Japanese consumer
is out of earshot and beyond the reach of the law of the French decision
maker, however powerful and determined he may be. The fall-out from nuclear energy depends in fact on a political and commercial meteorology
even more difficult to control than that which pushed the Chernobyl
clouds to unexpected territories. Faced with these overflows, which at the
same time as they give rise to identities, deeply wound them, the only solution is to go back to square one. The expression puts it well,[15] meaning that
we must revise the bases and rules of the calculation. We must count differently because we are counting with new actors who demand to exist as
such. How can we produce a common world in which there would be an
acceptable and recognized place for nuclear waste, knowing that deep
burial in the Gard may entail the disappearance of a population of viticulturists that nearly a century ago was prepared to shed its blood in defense
of its already threatened identity?

   As in the case of myopathy, the answer to this existential question can be
found only if we start from this new formulation of the problem. The question is no longer the one that, with its hint of irritation and peremptory,
formal tone, ends up reducing political decisions to the crude demands of
the bailiff: Something really has to be done about the waste, because it is

there, so what is to be done with it? It is, rather: What are the possible worlds in which nuclear waste could live together, in one form or another, with groups whose respected identities, whether of reserved farmers from the Meuse or stubborn viticulturists from Languedoc, could be recomposed in a collective to be devised? Hence, in order to get out of the impasse, the scientific and technological investigation must be re-launched, making sure that it is a co-piloted investigation in which the different actors concerned are not only associated with the definition of the problem, but also with the work of the research collective and the transposition of the results on the ground.

In the case of nuclear waste, the trajectory has not yet got so far as in the case of those suffering from myopathies, but some first adjustments seem to be taking place and it may well be that history is accelerating. These emergent groups who do want to give ground on their identity could be absorbed by bringing new technological options to light and by re-opening the question of the policy for energy. This is how those in charge at the CEA are beginning to recompose the research problems in order to take into account the (previously veiled) expression of these demands in the public space. In an interview given to *Le Monde* on 12 April 2000, the brand new general administrator, aware of the hostility of the population to the establishment of an experimental laboratory for the study of deep burial, begins by recalling the interest of classical secluded research: "It is worthwhile making the effort to find a site; the Americans, who have their site at Yucca Mountain, have discovered some interesting phenomena." After this call to order, he introduces some openings and new tracks that should enable the residents to be taken into account. Notably he emphasizes the interest of a solution, recommended by the 1991 Bataille law, of surface or sub-surface storage in very lightly buried sites. This route seems to him to be "very promising": "You reprocess the waste, store it, and you have access to it again if you want to reprocess it in the light of technological progress." The irreversible solution of deep burial, which entailed getting rid of the anxiety of the residents and of the waste at the same time, is replaced with a solution in which the voice of the residents is listened to and which leads to a re-examination of the options in order to reveal new ones (subsurface storage) that leaves the waste on hand and in view.

And, since several swallows are always needed to make spring, Yves Le Bars, also newly appointed president of ANDRA, after a career in which he had come up against questions concerning the management of the environment, drives the point home in the course of an interview given to the same newspaper. You talk of deep burial. But what in fact does this term mean? The question deserves our attention, since this adjective is at the heart of virulent protests, and besides it defines the option that it is the task of ANDRA to explore. There is clearly no question of him saying that investigations into deep burial will be abandoned, for this would mean disobeying the legislation. Yves Le Bars insists on this necessary discipline: "The law specifies that we develop a site for research on deep burial. But," he adds shrewdly, "there's deep and deep." Let's listen to him: "The fact of having accepted a granite outcrop gives a more supple margin to the definition than when one seeks a stratum of clay that is between 400 and 550 meters deep. The depth may start at 50 meters." For a long time we have hesitated between clay and granite for storing nuclear waste. The vicissitudes of geology have meant that clay is deep, while they have produced outcrops of the granite block at several spots. Now the task of ANDRA is to study geological sites that permit deep burial, that is to say, which provide nuclear waste with a burial place ensuring that risks of radioactive contamination are contained within the limits decided by legislation. On the contrary, surface or sub-surface storage means giving up reliance on geology for settling the problems of the waste and preferring to rely on the prudent vigilance of the populations. The granite outcrops permit two assurances to be combined: vigilance and geology. Thus, and no one had thought this before, we can now envisage getting rid of the waste without really getting rid of it, since we bury it while being able to reverse the decision at any time. The political game remains open, and it is by passing from the vertical axis to the horizontal axis of our diagram that the problem raised by the residents is unblocked, at least on paper.

In the case of sufferers from myopathy as in the case of nuclear waste, we can see a dynamic taking shape which enables the established and emergent groups to function on the two levels of the exploration of possible worlds and the composition of the collective. This dynamic, which is that of dialogic democracy, favors the exploration of problems, identities and the collective. But it is constantly under the threat of being absorbed too early by delegative democracy: emergent concerned groups generally do not have the resources to conduct this dual exploration; they are moreover easy prey for established groups which readily impede their actions when they consider it to be in their interests. If dialogic democracy is to be viable and not just wishful thinking, this dual exploration has to be framed. This twofold movement is conceivable only if research can be organized in a way to make use of both research in the wild and secluded research, and if specific identities can be expressed and debated. Thus it becomes possible to envisage a common world that is unexpected but compatible with the

objective results of research and with the production of subjective identities. The exploration that gives rise to an acceptable compromise develops
therefore in the space contained between the two independent axes of our
diagram. The axes open the field to a constant to and fro by which groups
are formed and change shape, composing their identities and the collectives to accommodate them. It is in this crucible that common worlds are
formed made up of mutant genes, bodies in pain, supervised nuclear waste,
viticulturists at peace with their Japanese clients, and sufferers from myopathy displaying their differences in the public space.

5   The Organization of Hybrid Forums

Faced with overflows which underline the inherent limits of the framings of delegative democracy, faced with the profusion of all sorts of hybrid forums (concerning the hole in the ozone layer, Bovine Spongiform Encephalopathy, myopathies, or nuclear waste), and faced with the ferment and continual turmoil that these introduce into our society, various measures of containment and channeling have been devised and tried out. For more than 30 years, in various places, in different modalities, and under different names, forms of organization have been tried and different methods tested whose single modest objective has been to introduce some rules of the game aiming to give some order to the conduct of debates and investigations.

These 30 years of experimentation have given rise to very few works of synthesis.[1] It is as if there is agreement about seeing these attempts as so many efforts of do-it-yourself without any general significance, or as occasional initiatives seeking to patch up a frayed delegative democracy. And yet, how can we fail to see that these experiments are serious attempts to establish new procedures and construct the bases for a deepening of democracy? Behind the hesitations and clumsiness, how can we fail to see the birth of a deeper movement, with the invention on the ground and by the actors themselves of original forms of consultation and deliberation? The diagram we have used to locate the new spaces uncovered by hybrid forums will continue to serve us to bring out, behind their obvious diversity, the unity of these attempts, and to reveal and make it possible to capitalize on the vicissitudes of experience that these attempts allow for. It will enable us to tackle the obviously central question of the influence of procedures on the double dynamic of the investigation of possible worlds and the composition of the collective. Rather than analyzing the philosophical or political science traditions, we prefer starting from practical experiences and the comments and reflections they give rise to and feed in to, to draw up a table of procedures that does justice to their diversity and specific effectiveness. Our approach starts from the problems encountered by the actors, it accompanies them in the analyses they produce, it follows the latter in the solutions they devise, and it strives to help them in the clarification of the lessons of more general import that may be drawn on the basis of the accumulation of experience.

When you go through the literature devoted to hybrid forums, you cannot fail to be struck by the agreement of all the authors on one of the major lessons that they draw from 30 years of experiments. "Wild" hybrid forums, those that no great effort has been made to discipline and organize, do not amount to simple agoras, to simple places of exchanges. There is nothing natural about their trajectories, the dynamic of which we have sketched out in broad lines in the cases of victims of myopathies and nuclear waste. They are the products of hidden struggles. The hybrid forum emerges at the cost of conflicts, often violent ones. To force a debate, and to be allowed to take part in it, you have to be able to call upon resources and put together alliances with a view to reversing the relations of domination that tend to repress any challenge to the double delegation. To leave hybrid forums to develop without any rules of the game for organizing the debate would leave the field free to the logic of relations of force, it would allow the reproduction, without discussion, of the exclusion of the weakest, precisely all those who seek to make their voices heard and be listened to. How many years will it have taken for the voices of those who seek in vain to express themselves on nuclear questions or on neuromuscular diseases to become audible? Established action groups that have been able to compromise with delegative democracy readily ally themselves with it to impede the emergence of concerned groups which could undermine their position. That is why dialogic democracy has to strive to strengthen the weak rather than weakening the strong.

But that's not all. Another way of getting rid of hybrid forums, without simply repressing the expressions of views that they allow for, is to instrumentalize them. This risk is underlined by all the authors, who are united in drawing attention to two frequent forms of manipulation. The first aims to use the hybrid forum as an apparatus for facilitating the drawing up of decisions that the decision makers sense are in danger of being debated at length. In order to anticipate unpredictable reactions, they find it a good idea to let people have their say, to give them the microphone, but having planned to turn it off once useful information has been obtained. The second is more cynical: The hybrid forum is reduced to a mere tool of legitimation. The decision makers consult, let people speak, but are careful not to take account of what is said and what it is proposed. In both cases people are given the chance to speak, but measures are taken to ensure that it makes no difference at all to the course of the decisions and that any attempt to organize the emergence of new identities is suppressed. In both cases it is a matter of getting people to speak in order to silence them more effectively, instead of flushing out the unexpected in what is said in order to give it weight.

When they finally see the light of day, at the end of violent confrontations that make it more difficult to organize a constructive debate, or when they are conceded by the decision makers with the un-avowed, but quite real and visible aim of anticipating objections the better to be able to brush them aside, or of giving the illusion of a debate, which they then ignore, "wild" hybrid forums do not bring any lasting contribution to the emergence of a dialogic democracy that would enable us to take the measure of the overflows which reveal the limits of delegative democracy. The tireless obsession of a number of actors has been to define and implement forms of organization that enable this enterprise of systematic sterilization to be avoided; the history of their attempts, involving stubborn patience and projects started over again a thousand times, should one day be traced.

How can we draw the lessons from 30 years of proliferating experiments? How can we make the inventory of the procedures that contribute toward the emergence of dialogic democracy? How can we evaluate the quality of the decisions that they enable? We will now try to bring some elements of an answer to these difficult but unavoidable questions.

But before undertaking this work of evaluation, we must start by drawing some boundaries by considering two procedures that have been and continue to be widely utilized. Both are situated on the edge of delegative democracy, aiming to compensate for its weaknesses but without giving overflows the space they demand, and this is why it is interesting to consider them together. These procedures are opinion polls and referenda.

The opinion poll is an instrument for identifying better the reasons why the public no longer has faith in the experts, and even entertains doubts about scientific and technological progress. Thus, since the 1980s, and on the initiative of public authorities and big multinational companies, many surveys have been undertaken to follow the evolution of opinion with regard to biotechnologies and to measure what was called their "degree of social acceptability."[2] Opinion surveys consist in questionnaires given to samples deemed to be representative of the whole population. The questions asked aim to assess, for example, the respondent's degree of optimism with respect to practical applications of biotechnologies, or to correlate these attitudes with social positions or levels of education. From these surveys it will be concluded, for example, that the majority of the public is anxious about the application of genetics to food, but that applications in matters concerning health are generally accepted. It will be established, moreover, that the more the public has full and rich information the more likely it is to support biotechnologies. Conclusions like the following may also be reached: "The most recent survey shows that the hysteria surrounding biotechnologies is not representative of public opinion."

Why not include these opinion polls in the universe of procedures that contribute to the organization of hybrid forums? First, because their explicit objective is to help develop strategies for getting the public of ordinary citizens, consumers, and more generally laypersons, to accept technologies or projects that the decision makers deem to be in the general interest, even though they arouse resistances that these same decision makers think are irrational. In this perspective, surveys make it possible to show that although they kick up a fuss, the recalcitrant are few, and to discover the points on which well-targeted supplementary information would promote acceptance.

The opinion poll reinforces the mechanisms by which delegative democracy protects itself against dialogic democracy. It is a procedure modeled on the electoral ballot. What counts, and what is counted, are individuals who are supposed to have personal opinions, which are framed by preformatting the questions and answers. The general will is extracted automatically by a procedure of statistical aggregation which is that of large numbers. Any possibility of constituting a space of discussion in which different identities and groups could emerge, which would make the question of the composition of the collective itself debatable, is carefully avoided. One of the most effective ways of preventing debate is to eliminate any possible link between scientific and technical contents, on the one hand, and the composition of the collective, on the other. The two dimensions, the establishment of a relation between which is, as we have seen, the cornerstone of hybrid forums, are separated from each other. Finally, the survey results in a complete reification of public opinion, which is summed up in a few propositions that can be used without the consent of the public itself, which is dispossessed of control over its own opinion: "As the euro-biotechnology barometer has demonstrated, the public has given proof of its maturity by answering that it is ready to accept the controlled use of GMO in the food chain." The public has nothing more to say and cannot comment on what it has been made to say. The only thing that counts is the "opinion" that has been produced, and of which the public is dispossessed once it has been gathered. In some cases this opinion is strongly suggested by the communication strategies of industrial groups that publicize the answers to the questions that are asked.

When it aims to prepare for a choice, the referendum merely reproduces the opinion poll, but on a larger scale. The only difference, albeit a weighty one, between surveys aiming to get the public's attitude toward genetically modified organisms, and the Swiss "vote" to determine whether or not research on GMO should be banned, is the direct link between the vote and the political decision. Even if the referendum does not determine the decision mechanically, the fact of its solemn character means that it weighs on it. But apart from this difference, which is not negligible, the logic and principles are the same.

Like the survey, the referendum is addressed to individuals assumed to have preferences that they know or are capable of expressing when asked the right questions. The referendum, like the survey, preserves the double delegation; it even helps to reinforce it. In fact, the question never bears on the possibility of collaboration between research in the wild and secluded research, but on the interest of the latter. Furthermore, the ordinary citizen is given the chance to have a say only for this to be immediately withdrawn, leaving him no other initiative than to tick the box corresponding to the answer he wants to give to a question he had not really chosen.

At the same time as the referendum reinforces the double delegation, it plainly expresses its limits and contradictions. By withdrawing the monopoly on decisions from the legitimate representatives and specialists, it underlines the impotence of delegative democracy in the face of cases that plunge decision makers into uncertainty; but, in addressing itself to the ordinary citizen, it prevents the formation of groups, the emergence of identities to be discussed, or, in a word, the exploration of a common world. Under these conditions, as the Swiss vote on GMO demonstrates, the referendum is a bit like a game of Russian roulette. To scientific and political uncertainties, which it leaves unresolved instead of trying to take them up, the referendum adds the irreversibility of a decision made in complete ignorance of the facts. The sovereign people that the referendum puts on the stage is a people that has been deprived of any capacity for investigation and the gradual and active search for compromise. We ask it to decide, trapping it in the rigid frameworks of delegative democracy, but deny it the possibility of reworking questions calling for further information. Politics is reduced to a caricature of itself. A few votes would suffice for the Swiss to ban GMOs for ever, just as a few votes would suffice to commit the Swedes to a program of nuclear energy.[3] Who can believe that the construction of a common world can be based on such procedures? It is not the people who are irrational, any more than it is the representatives or specialists, it is this delegation that does the rounds, which no one wants, and that the actors, at a loss at what to do about it, enthusiastically pass on to each other.

If procedures are to contribute to the emergence of a dialogic democracy they must break down the monopoly of the double delegation, if only at the margins. As we have just seen, opinion polls and referenda aim instead to preserve this double monopoly. While handing back the voice to ordinary citizens and laypersons, they maintain the two breaks on which delegative democracy is founded. Surveys and referendums cannot therefore be included in the set of dialogic procedures.

Criteria for Classifying Dialogic Procedures

Organizational Criteria

Having rejected some of the procedures that try to save delegative democracy, we now need to sort out those that contribute to dialogic democracy. In order to do this, a useful starting point will be the diagram we used to map out the space in which dialogic democracy develops. (See figure 4.3.)

The different procedures can be evaluated in terms of their ability to facilitate a deepening of the democratic regime and so go beyond the limits imposed by respect for the double delegation. Two criteria, applicable to both the axes, that is to say, to the two forms of delegation, will enable us to construct a grid for evaluation taking account of the different degrees or levels of democratization introduced by the procedures considered.

The first criterion measures the distance covered along the axis being considered. What is the *intensity* or, if one prefers, the *depth* of the challenge to the divide imposed by each of the two delegations?

With regard to the production of knowledge and the exploration of possible worlds, the procedures are characterized by the way in which they enable non-specialists to collaborate with specialists and whether or not they create the possibility for close and strong cooperation between secluded research and research in the wild. The intensity of this collaboration is easily measured by how *early* laypersons are involved in research. A procedure contributes to a greater or lesser extent to overcoming the division between laboratory research and research in the wild according to whether it affects the identification and formulation of problems, the extension and organization of the research collective, or the application of laboratory results in the real world. Thus, some procedures promote the participation of non-specialists at the point of formulation of the research problems, while others introduce them at the end of the process, when it is a question of transposing and adapting laboratory results on the ground. The earlier cooperation between research in the wild and laboratory research is organized, the more the procedure is likely to affect the three operations of translation.

On the second axis, procedures are distinguished in terms of their capacity to foster the composition of the collective and make concern for the collective an imperious preoccupation. Some are content to facilitate the assertion of emergent identities, providing the possibility for "supporters of small causes" to enter the public space and make their voices and differences heard. Others go further, organizing an early exchange between emergent minorities and even established groups who are encouraged to listen to one another and reach agreement. Others, finally, push the groups to negotiate and enter into the composition of the collective by making the most of the instability of identities and the adjustments this allows for. Thus there will be greater or lesser concern for the collective depending on whether or not the procedure goes beyond just the assertion of the emergent identity and takes the mutual listening to and even negotiation of identities into consideration.

Whether it is a matter of the exploration of possible worlds, and so of the organization of research activities with a view to the production of new knowledge, or of the composition of the collective, in both cases the number and diversity of the groups which are mobilized, and so concerned by the debate, provides a further criterion for assessing the degree of dialogic democracy allowed for by the procedure in question. This is, thus, a criterion of *openness*: To what extent are new groups invited to express their views, exchange their points of view, and negotiate? The more groups there are and the greater their diversity, the more meaningful the debate will be. The criterion of openness enables us to distinguish between procedures that restrict access and those that, on the contrary, enlarge it. They apply in the same way to both of the axes, that is to say to each of the two forms of delegation. What groups are encouraged to take part in the dynamic of cooperative research and the composition of the collective (whatever the intensity of participation)? To what extent will the groups with access to these two dynamics have the power to modify their identity and expectations as a result of the scientific or political investigations taking place? The assessment of the degree of openness of a procedure depends on the answers it gives to each of these two questions. Two more specific criteria make it possible to account for this. The first directly measures the openness made possible by the procedure in relation to constituted and visible action groups by taking into account the degree of *diversity* and *autonomy* of the groups mobilized. The second concerns the greater or lesser ability of the procedure to allow for and encourage the repeated redefinition of emergent identities. What is in question is the ability to follow the transformation of the groups, to take it into account, and, consequently, to leave the question of the *representativity* of the spokespersons who speak in the name of their constituents open and debatable.

Whatever the intensity and openness of the debates that a procedure allows for between laypersons and specialists, or between representatives and those they represent, the *quality* of the collaborations and discussions it is likely to encourage is itself variable. How far can emergent identities go in their presentation of themselves, up to what point can mutual knowledge be deepened, and what ability in argument and counter-argument do the groups have when they engage in the composition of the collective? But also, how far can research in the wild and secluded research push the discussion of problems, debate the boundaries of the research collective, and be involved in the adaptation of knowledge?

This quality, for both axes, is assessed from a double point of view. The first is associated with what could be called the *seriousness* of voice (are the protagonists able to deploy their arguments and claims, as well as answer objections, with the requisite acuteness and relevance?), while the second provides a measure of the degree of *continuity* of voice (are the interventions and discussions spasmodic or can they last?)

Table 5.1 gives a synthetic presentation of the three criteria (intensity, openness, quality) and the six sub-criteria which will serve to classify the

Table 5.1
Degree of dialogism of procedures.

Criterion	Sub-criteria	Value	
Intensity	Degree of earliness of involvement of laypersons in exploration of possible worlds	Strong	Weak
	Degree of intensity of concern for composition of collective		
Openness	Degree of diversity of groups consulted and degree of their independence vis-à-vis established action groups	Strong	Weak
	Degree of control of representativity of spokespersons of groups involved in debate		
Quality	Degree of seriousness of voice	Strong	Weak
	Degree of continuity of voice		

procedures in terms of the degree to which they contribute to the establishment of democratic confrontation. Procedures will be deemed to be more dialogic to the extent that they encourage exchanges and debates that are intense, open, and of quality. Thus, through an enrichment of the previous diagram, what can be called the *normatively orientated space of dialogic procedures* will be defined.

It will have been noted that two sub-criteria correspond to the first criteria (intensity), specifying the meaning it assumes with respect to the challenge to each of the two delegations. The two other criteria (openness and quality) and the four corresponding sub-criteria apply both to relations between laypersons and specialists and to the relations between representatives and those they represent, since groups are active on both the axis of the production of knowledge and on the axis of the composition of the collective.

Implementation Criteria

To these six sub-criteria, which provide a grid for evaluating the contribution of procedures to democratic confrontation and dialogue, must be added criteria for assessing the conditions of implementation of these procedures.

All debate is actually permeated by asymmetries, generally transmitted and reinforced by the procedures of delegative democracy. Established action groups and their accredited representatives, as well as elected representatives, tend to monopolize discussion; secluded research tends to exclude research in the wild. According to a logic of reproduction, backup forces are deployed to confine the overflows and the uncertainties they bring along with them, so as to contain them within the space of delegative democracy at any cost. The deployment of the dialogic space in accordance with the six sub-criteria of table 5.1 becomes more problematic and difficult as one moves further away from traditional procedures in order to involve research in the wild earlier, intensify concern for the collective, increase the representativity, diversity, and independence of groups, and improve the seriousness and continuity of voice. If weak voices are to be able to make themselves heard, and as soon as possible, if they are to be given the possibility of playing an active part in the composition of the collective, and if they are to be listened to and be influential, then they need to be assured of the resources of time and money, as well as training.

Both the nature and the volume of the necessary strategic resources depend on the importance of the challenge to the model of the double delegation. For example, it is more difficult to participate in the identification and formulation of problems than to wait for researchers to come out of their laboratories. Involvement in debates specifically concerning the composition of the collective and the adjustment of identities is no less demanding in terms of skills and know-how. Think of the difficulties that patients' associations encounter when they strive to maintain their specific demands while showing the place they could occupy in a reconfigured collective. To get the debate going and make their points of view intelligible and debatable, they have to pass from a formulation of the problem limited to particular diseases to broader and broader formulations, which integrate increasingly large populations, in order eventually to be able to ascend still further in generality by putting forward a re-translation of their demands into more abstract and universal notions such as those of the human person and human dignity. We have suggested that this link takes place by calling upon collaborative research (and therefore through access to specific resources), but also through consultation with patients as well as through the acquisition of a socially rare skill that enables unprepared actors to conduct a discourse of great generality (and so ambition) in public arenas. In a very interesting work on the Comité consultatif national d'éthique (the National Ethics Committee, set up by François Mitterrand in 1983 to make recommendations to government authorities on bioethical issues), Dominique Memmi shows that the ability to put concrete patients out of mind so as to speak only "about" the patient in general, and then of the rights of the human person, presupposes preparation in terms of education, indeed in terms of professional skills, which are far from being equitably distributed in the population.[4] To remedy these asymmetries it may be useful to envisage the formation of new professional roles: translators, mediators, facilitators of debates and negotiations, and political organizers whose explicit task would be to make it easier for previously excluded actors to enter the public space. Thus the costs linked to the establishment of *equal access to the procedure* must be evaluated, *ex ante*, and resources must be released to cover them. If this condition is not met, the best procedure in the world will quickly be transformed into a masquerade and an enterprise of collective mystification. Only already well-established actors with the ability to make themselves heard and understood will participate, those who, accustomed to dominant positions, are quite capable of being haughty in defending their points of view. Instead of being widened and enriched, debate becomes narrowly confined to those who already have a monopoly on legitimate voice.

The *transparency* of the procedure is a second criterion that enables its implementation to be assured and controlled. How can we know who has made a contribution? How can we preserve the record of the positions taken? How can we establish the ordered sequence of the arguments and counter-arguments developed? In a word, how can we reconstruct the dialogical richness of a debate if there are no recording tools making it possible to track the different voices? The very notion of a public space or of interrelated public spaces presupposes this process of making visible. The meaning of the notion of transparency should not be misunderstood. In no way does it entail revealing the whole of a social body hidden by a veil that the procedure tears away, a world that was already there, demanding only recognition. Transparency applies only to the procedures themselves and the way in which they structure and organize the public space. The latter, simply through its existence, expels a whole set of actors, problems, and question into the shadow of the private sphere. Transparency always has a high price, which is the opacity that is its corollary, but the worst situation is one in which, to the inevitable but always re-negotiable exclusion of actors and causes relegated to the shadows of the private sphere, there is added the more insidious exclusion of those actors who play a part in the public space but whose voice is lost. Transparency may also involve having recourse to a judge when procedures are not respected.

Transparency must not be only retrospective; it equally applies to the future. To avoid manipulations, which necessarily benefit the strongest, the procedure, and the different actions and operations in which it is concretized, must be known in advance by all the participants. This third criterion is that of the *clarity (and publicity) of the rules of the game.* Agreement on how to proceed should leave no point unclear and, once obtained, is a firm commitment; there should be no question of going back on the rules agreed by the different parties involved.

Table 5.2 brings together the criteria just presented.

The Procedures

We now have a battery of criteria that will enable us to classify procedures in terms of their ability to foster dialogic democracy, or in terms of what we

Table 5.2
The implementation of procedures.

Criterion	Value	
Equality of conditions of access to debates	High	Weak
Transparency and traceability of debates	High	Weak
Clarity of rules organizing debates	High	Weak

can call their degree of dialogism. A procedure will be more dialogic (that is to say, it will facilitate the double exploration of possible worlds and identities, and of the collective to a greater extent) when it corresponds to strong or high values of the different criteria in tables 5.1 and 5.2.

A great wealth and variety of procedures have been devised since 1970. Furthermore they have developed and evolved. The actors who devised and implemented them have in fact benefited from the experiments undertaken. They have been quick to change procedures in the light of the lessons of experience, combining them with each other and tinkering with them to make them better adapted. Under these conditions there is no point in trying to draw up an exhaustive list. It is impossible to do justice to this wealth and variety. Our objective is more modest and limited. We would like to show the operational character of the criteria of evaluation we have proposed while limiting ourselves to a small number of procedures taken to be representative of the diversity of those that exist. The obviously central question of the connection between procedures and decision-making processes, or rather, of the integration of procedures within a dynamic of construction of the public space, will be left to the following section.

The order of presentation we have adopted has led us to rank the procedures in terms of their degree of dialogism. That is why we will begin with *focus groups* and end with consensus conferences. To avoid wearying the reader, we will merely review all the criteria for the consensus conferences. Because the names designating the procedures are rarely stabilized and standardized, the names we have chosen are somewhat arbitrary.

"Focus Groups" or Discussion Groups

The origin of focus groups goes back to the Second World War and the efforts of the producers of propaganda films to assess their impact.[5] They had the idea of organizing projections for differently composed audiences in order to evaluate their effects. Spectators were asked to press a green button when they supported the message and a red one when they felt they were offended. The projection was followed by a discussion, with a leader, so that reactions and comments could be expressed. This method then became very popular in marketing as a way to identify consumer preferences and tastes. In the case of the food-processing industry the practice takes the form of groups of consumers who taste the products (wines, cheeses, soft drinks...) and who are asked to communicate their judgments and evaluations in a form that can be used by the designers or specialists of commercial promotion. *Focus groups* have also been used in the field of health, for gauging the effectiveness of AIDS information campaigns, for example, or for testing the influence of family planning on contraception practices. They have become common practice in the political domain. In recent years the procedure has been used increasingly often to explore the attitudes and expectations of the public on subjects concerning the environment, or the social acceptability of technologies more generally.

The suppleness of the device is due to the fact that several groups can be formed, each of them containing from five to a dozen persons, so as to take better account of the diversity and heterogeneity of the public. On a health problem, and more precisely on the choice of drugs, for example, three groups could be formed, the first bringing together professionals, the second containing patients, and the third consisting of representatives of the pharmaceutical industry. Each group holds only a few meetings; when the subjects are not too complex, one meeting usually suffices. On the basis of very open and very general questions asked by the leader, a free discussion is set up which is recorded and sometimes filmed. In comparison with questionnaires or opinion polls, *focus groups* allow for the deployment of a collective dynamic and the possibility of positions emerging that could not be included on any list of closed questions. The procedure has been mobilized on a number of occasions in cases concerning science and technology, as when trying to get a better grasp of local residents' perceptions of an incinerator, or when reflecting on national energy policies.

Paradoxically, such a widely employed procedure (more than 100,000 focus groups are organized annually in the United States alone) has aroused hardly any interest in the social sciences, which prefer great doctrinal debates to the fine analysis of procedures invented by actors. The attempts at synthesis put forward by the practitioners pick out the following lessons:

- For a focus group to work effectively, it seems that on average it should have around six members.
- Recruitment generally takes place by telephone, following a procedure of random selection.
- Each session lasts about 2 hours.
- Sessions are held weekly and the groups meet in a neutral place.
- The use of audiovisual material creates an emotional state favorable to the expression of opinions.
- The products of the group activity are audiovisual recordings, answers to questionnaires, and even written reports when assurance has been given that they will be taken up in the policy decision process.

- Discussion is facilitated by a leader (*animateur*), who must not be an expert.
- A session of six persons costs about US$1,000.

Furthermore, focus groups are seen as being able to make a positive contribution to technological evaluation.[6] The rapid application of the grid of evaluation enables us to clarify the nature and significance of this contribution.

Let us acknowledge first of all that the contribution of this procedure to the establishment of a dialogue between research in the wild and secluded research is very marginal. Nevertheless, it can stimulate this dialogue by bringing to light new tracks or more often by encouraging an awareness of the need to change the hierarchy of themes already being investigated. In reality, its contribution consists for the most part in initiating the expression of expectations, of still inchoate interests, which are thus able to take shape and be endorsed by those putting them forward. If the exploration of diversity may be fairly wide and rich, it nevertheless always remains on the threshold of the construction of collective identities. Even if a fine work of analysis sometimes allows a glimpse of identities in the process of formation, it has to be acknowledged that the voices heard in focus groups rarely leave the universe of tried and tested established identities, which ask only to be taken into consideration, or only to express themselves on the subject under consideration. This explains why the designation of spokespersons, with its procession of consultations and discussions, is never broached. Let us add that focus groups lack both continuity and seriousness of discussion, which, through its construction, is necessarily occasional and superficial.

Let us now consider the criteria concerning the conditions of implementation of the procedure. The assessment is mixed, although positive overall. Access to resources is, to say the least, problematic; members have very little time available for the elaboration of their points of view, and training, which is fairly frequent when the case is complex, nevertheless remains very rudimentary; traceability is good since recording is the rule; clarity of the tasks is also high.

Public Inquiries

Focus groups are widely used, but the conditions in which this is done and the modalities of their implementation vary considerably. Other consultation and political decision-making procedures are by comparison more formalized. We are now going to examine these, starting with a set of tools which can be grouped together under the generic term "public inquiries," and which are found in fairly similar forms in various countries. We will start with the French case to present their functioning and assess their contribution to the establishment of dialogic democracy. This evaluation exercise could however easily be transposed to other countries.

Public inquiries aim to reconcile two different objectives: (1) the effectiveness and safety of public decisions through some degree of transparency regarding the reasons and contents of the projects and (2) recognition of the right of populations concerned by a project to express their views, and even to object to it. These two distinct objectives are supposed to combine to produce a social acceptability that avoids local conflicts and prevents disputes.

At the origin of French public inquiries we find the "declaration of public utility" procedure created in 1834 to protect the economic interests of private landowners when the state expropriated their land for civil engineering works (roads, bridges, etc.). But from the 1970s numerous local conflicts and mounting opposition to this type of infrastructure resulted in changes to the procedure. First, the obligation to inform the population on such projects was introduced (1976), followed by the official recording of the public's observations (1983). Then, for major projects (high-tension electricity lines, highways, airports, etc.), a national commission for public debate (CNDP—Commission nationale du débat public) was set up (1995) to organize more in-depth consultation with people living in the vicinity.[7] This commission adopted the methods validated in North America, especially those of Quebec's BAPE (Bureau des audiences publiques pour l'environnement).[8] Between 10,000 and 15,000 projects are examined via the general procedure annually, and some 30 via that of the CNDP.

Many observers in the social sciences identify two recurrent problems. The first is the very weak public participation in inquiries that lack preparation. It is often said that "public inquiry = inquiry without a public." The second problem is the fact that these procedures have a particularly weak impact on decisions. Only about 5 percent of all opinions expressed are negative, and consultations rarely seem to have enriched or altered the project under consideration. In practice, therefore, the public inquiry is not a tool of consultation but one designed to gain adherence to a project.

These limits should not hide the effects of the procedure on the organization of hybrid forums. In the first place, at least in France, this procedure remains the only obligatory moment of public consultation for thousands of projects. This explains why, with increasing frequency, forces spontaneously emerge to counterbalance the natural tendency toward a restrictive orientation of this consultation. The mobilization of a wide public may be carried out through existing social networks (families, neighbors, friends), as well as by organized groups that arouse participation by deploying mechanisms intended to facilitate access to the procedure (e.g., by circulating observations submitted in the requisite form). Some associations sometimes go so far as to offer to go to the municipal offices (where the documents to fill in are kept) for the signatories. "They thus enable the population to participate without moving, or even picking up a pen, except to write a name and address."[9] This mobilization can prove to be highly effective in pressurizing the elected representatives, on whom a public inquiry (fixed by decree) is imposed without their having decided it. The same applies to the experts who are appointed as inquiry commissioners (and who are generally engineers, architects, or retired magistrates).[10]

We know enough about the theoretical and practical functioning of the public inquiry to evaluate the procedure with the help of the set of criteria we have proposed. There is undoubtedly more encouragement for cooperation between secluded research and research in the wild the more the public inquiry and the debates to which it gives rise move upstream of the projects. As for concern for the collective, this depends on the practical conditions of the procedure's implementation. It may not amount to much when the only interests expressed are those of the residents (the notion of resident being well suited for designating groups directly affected by any kind of project). If the procedure welcomes associations, and if the project itself affects a multiplicity of places, sites, or different populations without being broken up into separate and independent parts, then the process of the composition of the collective may get underway, for the inquiry can go together with the exploration of identities and forms of knowledge that are not necessarily limited to the residents *stricto sensu*. In its principles at least, the procedure is sufficiently open to facilitate snowball effects that may gradually lead initially distant groups that are *a priori* indifferent to each other, but which are increasingly concerned by the case, to join forces. At the end of the process it may be the French as a whole who see themselves as the residents of the future site and who, as a result, feel entitled to express their point of view. Seriousness and continuity of exchanges may be obtained by setting up durable structures of discussion and having a manned office or by inviting experts according to the different subjects being treated. To summarize, and without underrating the practical limits that all the analysts acknowledge, the public inquiry, through its structure, and especially if influenced by determined wills, may lead to the organization of broad and open hybrid forums. From this point of view, it will have been noted that the implementation criteria are particularly important.

Consensus Conferences

This procedure, which appeared in Denmark (in which eighteen consensus conferences had been organized by January 2000), has been implemented by a number of other countries on every continent (it has been used in fourteen countries as different as Japan, the United States, the United Kingdom, Holland, New Zealand, Canada, and ... France).[11] The subjects considered are linked to current events and include a strong scientific and technical component. They are generally characterized by the existence of a well-established professional expertise. This does not prevent them from giving rise to heated controversies that are the consequence of considerable uncertainty about the possible effects of implementing the technologies being questioned. Consensus conferences, which some countries have renamed—France preferring to call them "citizens' conferences" and Switzerland "publiforums"—were devised in order to include the "public" in the circle of discussion usually limited to decision makers and experts. Their explicit aim therefore is to bring about and structure the widest possible debate with a view to enlightening decision makers on technical issues about which there is still considerable uncertainty. To present the procedure and its operation we will take the example of the French citizens' conference organized in 1998. It concerned the use of genetically modified organisms in agriculture and the food industry.[12]

When the prime minister announces this decision, the case is both confused and much debated. The government has in fact just authorized the firm Novartis to cultivate transgenic corn, challenging, without prior consultation, the prohibition pronounced by the previous government. The latter's policy, moreover, had not been very clear, since when it declared the prohibition it was authorizing the import of transgenic corn!

The French Parliamentary Office for the Evaluation of Scientific and Technological Choices, which is given responsibility for organizing the conference, will not try to innovate. It will be content with applying a tried and tested model. It forms a panel of fifteen citizens (seven women and eight men) who are selected by an opinion survey institute. These citizens are chosen randomly and in such a way as to be representative of civil society to some extent (there will be managers, farmers, employees, and so forth on the panel). They are genuine laypersons with no knowledge of the case and not directly affected by the decisions at stake. They are selected so as to ensure the greatest possible diversity of opinions. When it has been formed, the panel undergoes training sessions in which academic specialists take turns presenting them with the knowledge they will need to grasp the content and significance of the debates taking place. In the French case, this training is provided over two weekends. The following themes are considered:

- the development of agricultural production in recent years
- industrial techniques of food processing
- general principles of nutrition
- basic facts of genetics
- improvement of vegetable species and transgenesis.

The first weekend is devoted to presentations and the transmission of knowledge and information. Over the second weekend, members of the panel, who were happy to be attentive students, are asked to transform themselves into citizens concerned with the common good and to raise questions about the issues and problems raised by GMO. The following subjects are dealt with in this way: the national and international legal context, the environmental issues, the agricultural issues, and the food-processing issues. At the end of this second weekend the group works out a grid of questions that will structure meetings with the experts. The following stage is actually the conference itself, during which these simple citizens question the experts and enter into dialogue with them. In the French case, the experts are chosen by the group with the help of the steering committee which, from the first moment, acts as methodological adviser to the Parliamentary Office.

But to speak of experts is to misuse language. More exactly, they are spokespersons of established groups who over the years have developed a good knowledge of the issue and, to a greater or lesser extent, have been involved in the public space of discussion and the controversies aroused by GMO. It would no doubt be more accurate to see them as parties to the debate, those who, owing to their abilities, convictions, or interests, are directly affected by and involved in the issue. The English speakers use the word 'stakeholders' to designate these groups. They include industrialists, more or less senior civil servants, representatives of consumer associations, leaders of professional associations, spokespersons of non-governmental organizations, like Greenpeace, and even representatives of political parties. They also include specialist academic researchers into GMO who, owing to their abilities, tend to hold strong but nonetheless informed opinions. In a word, included among the experts are all those who, within the framework of delegative democracy, had the ability and know-how to form interest and issue groups that have at some time put pressure on decision makers to take measures in line with their interests or convictions.

Then there is the conference itself, that is to say the organized dialogue between the members of the panel and the experts. In the French case this takes place in public in the premises of the National Assembly. The media are there; journalists interview members of the panel during pauses in the proceedings; the public take part. The sessions are directed by the president of the Parliamentary Office for the evaluation of scientific and technological choices. On each theme selected by the group of laypersons, four to six experts (in all, 27 will take part) make a short presentation of five minutes and then answer questions.

Once the conference is finished, the group of citizens withdraws and in a very short time drafts a document, a written opinion, which forms the answer to the question it was asked. In the French case, this work is carried out in an afternoon and a night. It is presented at a press conference. As is noted by members of the steering committee (which is made up of seven people, all academics, and which meets no less than fourteen times before the final conference): "The calm apposite way with which each succeeds in facing the questions creates an atmosphere of proud modesty and shared honesty that is felt with real emotion by the many participants, including journalists."[13]

The procedure, in the form we have just described, apart from some minor adaptations and variations, is the one followed in most of the experiments undertaken until now. In every case we find the steering committee, the group of citizens, the training sessions, the public dialogue, and the drafting of a document in the form of an opinion. What varies are the rules for designating the citizens—although these are always based on a random selection—the concrete modalities of the training, the existence or non-existence of a document written by the organizers before the conference setting out a list of questions on which the panel's opinion is formally requested, and the duration of the procedure. What also varies, but we will come back later to this point that deserves our full attention, is the place of the conference in the policy decision-making process.

Apart from these differences, consensus conferences play an almost identical role in the organization of the public space. To describe it, let's return to table 5.1 and the different criteria put forward. Being a very ambitious, as well as a very popular procedure, we will undertake a detailed evaluation in order to provide a concrete example of the use of our grid of criteria.

Intensity? The consensus conference is not content with recording scientific facts. It takes up a position at the heart of uncertainty and the research process strictly speaking. In the case of the French conference, the citizens show themselves to be keenly aware of these uncertainties and one of the group's recommendations aims to set up an organization of research that takes them into account. Of course, one of the obvious limits of the procedure is that the conference cannot go so far as to put a form of collaboration between research in the wild and secluded research into practice. What it can do, and does, is to plead for better coordination between these two modes of research. Thus the panel lays great emphasis on procedures of bio-vigilance and the evaluation of experimentations, as well as on the need to establish traceability of GMO. The explicit aim of these rules, whose implementation is strongly recommended, is to open up, vascularize, and let some air into secluded research by introducing new actors. We can say therefore that the consensus conference takes a step in the direction of establishing the beginnings of formal collaboration between research in the wild and secluded research. This step is all the more credible in that in the debates the ordinary citizens have demonstrated not only what we are happy to call their wisdom, but also their ability to grasp the strategic dimensions of research. Some conferences have gone further, like the conference on cloning organized in Holland, which recommended launching a social science research program on the links between cloning and identity.

As regards concern for the composition of the collective, we should note that the known experiments of consensus conferences lead to a more critical appraisal. Citizen panels generally manage, with acknowledged success, to distance themselves from established interests. As the French experience with GMOs shows, this distancing is built into the procedure, due to random selection, the absence of conflicts of interest for members of the panel, and the organization of hearings in which different interests are put on the same level; each group, whether it is worth billions of dollars or a few thousands, whether it brings in millions of votes or just a few, is entitled to the same time to speak. Thus one of the central recommendations of the panel was to ban marker genes of resistance to antibiotics, but also to reform the Bio-molecular Engineering Commission (CGB), which advises the government, notably on the question of GMO, so that its relation to established interests is less that of follow-my-leader. In both cases the group of citizens proposed measures aiming to favor the collective, something of the order of the long-term general interest, in comparison with the natural play of particular interests and the short term. As some observers emphasize, "Laypersons bring a vision freed from localized stakes and this enables general concerns linked to the control of technology in society to be reintegrated into the analysis."[14] The consideration of global stakes, that is to say, those concerning the collective as a whole, tends to be focused on questions of risks and responsibility, but sometimes it may go so far as to include ethical or socio-political questions. This step was not taken in the case of the French conference on GMO. According to the observers, this shortcoming can be imputed to the haste in which the conference was prepared and held (five months in all, whereas foreign experience shows that a full year is needed), as well as to the lack of previous experience on these subjects. However, in other countries and on different subjects, ethical or political questions have been taken up. In the French case, one might have thought, for example, that the effects of the development of GMO cultures on inequalities between North and South, as well as on agricultural development, or the more general theme of the "patentability" of the living, would be discussed and give rise to some recommendations emphasizing the interest of protective measures to allow time for a more thorough debate to take place. This would have been all the easier as there were already public positions on all these subjects.

Be that as it may, whether it is a matter of economic, political, or ethical interests, the group of ordinary citizens confines itself to a traditional vision of the collective and the general will. The latter is not arrived at through the composition and adjustment of emergent identities. The abstract individual remains the basic element on which the collective is instituted. The panel alternates between considerations concerning individuals (protection of the consumer, the right of individuals to express their views in the face of pressure groups) and questions affecting the collective interest (allowing choice between several types of cultures and food-processing industries, taking care of future generations), but without turning its attention to the formation of middle-range groups striving to express a still unrecognized emergent identity (like the movements that appear on this occasion, appealing to different modes of organization of the world market and challenging the power of certain multinationals). The citizens' conference prevents more than it facilitates the organized discussion of these positions.

The citizens' conference helps call into question, at least programmatically, the break between secluded research and research in the wild to the same extent as it fails to give greater prominence to concern for the uncertain composition of the collective. This limit is inscribed in the procedures. The political aim of the consensus conference is to make a systematic inventory of, and make audible, constituted points of view, some of which either cannot or do not want to make themselves heard in the public space.

Its specific effectiveness is that of passing from the obscure logic of pressure groups to one of positions taken in a space where every voice has equal worth. But rendering points of view visible in this way does not have the objective of getting a dynamic of exploration of the collective underway; its basic aim is to reintroduce into the mechanism of delegation points of view that were not overtly taken into consideration. Before the consensus conference, a party could ignore the question of GMO and refrain from taking a position; the ordinary citizen was thus deprived of the possibility of choosing between different constituted arguments. After the conference, it becomes more difficult to maintain this abstention. Citizens may require candidates for delegation to express themselves on the subject. As we see in this example, the citizens' conference does not open up the debate in order to facilitate the emergence of new identities, or in order to make new voices heard by offering the possibility of getting exchanges going with other voices with a view to putting the constitution of the collective up for discussion (how can we organize markets? what is the future for agriculture and farmers?), but it opens the debate so as to force candidates for delegation to propose to their electoral clienteles public positions and programs of action that integrate the question of GMO. The consensus conference does not challenge political delegation; it aims to make it more effective, but without affecting the break between ordinary citizens and their representatives.

Openness?   If the citizens' conference does not allow for the dynamic and interactive expression of new identities, it does make it possible to take the measure of the greater or lesser popularity in the population of already established and articulated positions. The procedure contributes to the public assessment of existing positions which are inscribed in a spatio-temporal framework sufficiently narrow for their comparison to be possible, and indeed unavoidable. As a result, the panel is put in position to judge between the arguments. Furthermore, it is enjoined to make judgments and evaluations. Confronted, physically we could say, with the great diversity of already formed interests, it is led to detach itself from them. Relativization is not a mental disposition peculiar to laypersons but the consequence of the procedure adopted. From this point of view, whether it is a question of the production of knowledge or discussion of the general will, the consensus conference constitutes a fairly effective opening set of arrangements which allows for an objective inventory of positions and facilitates their expression in the public space.

As a general rule, it is difficult to challenge the representativity of the spokespersons since the procedure favors well-established groups. When the director of research from Novartis expresses himself, when representatives of the Greens or of Greenpeace answer the panel's questions, there is no doubt that they are committing their constituents. But, we repeat again, at no time is the possibility opened up of a dialogue between a group in the process of being constituted and those who, for a time, are designated as its spokespersons. This work of iteration, which leads to a more or less rapid turnover of representatives and allows the formation of an identity through trial and error, is obviously excluded by this procedure.

Quality? The general view of observers is that there is no doubt about the seriousness of the exchanges. Whether discussing amongst themselves or with the experts, members of the panel express themselves while taking their time and listening. They ask questions that aim to increase understanding of the problems. Furthermore, and this has also been often underlined, they let their convictions and their emotions speak without the restraint of any censorship. If these conferences are moving, it is because they preserve the emotional dimension of all public debate, even when it is framed by rules aiming to make it reasonable. By neutralizing calculation, the procedure produces both authenticity and good faith.

The continuity of the debates is obviously very weak. The procedure condenses discussion and exchanges into a strictly framed time and space.

It remains to continue the analysis by reviewing the criteria linked to *implementation* of the procedure. Access to resources is not marked by any flagrant inequality. Members of the panel do not encounter any particular financial restraints; they have access to training and to the media that interview them and give important coverage to their proposals. The discussions and, more generally, the whole procedure are traceable, since everything is recorded and filmed, and these recordings are accessible to all and sundry. Transparency could easily be improved, by allowing the conference to be followed in real time on a television channel for example. Finally, how the procedure works and the roles of the different participants are clearly defined *ex ante*.

This examination has shown the operational character of the grid of criteria we have proposed for assessing the degree to which the procedure contributes to dialogic democracy, while emphasizing both its contributions and its limits. We have chosen to apply this grid to consensus conferences because this procedure is one of the most popular and well known, and certainly not because we consider that it is the most dialogical of all. The consensus conference is an effective tool for making a meaningful start on recognition of the possible role of laypersons in scientific affairs; it also enables the foundations of delegative democracy to be improved by making the pressures of the most influential established action groups both visible and debatable and by opening the public space to those who were excluded from it. The other side of this is that the exercise does not allow a real exploration and formation of new identities, or the composition of the collective that could result from this. Nor does it contribute to the concrete organization of collaboration between secluded research and research in the wild, although it facilitates its eventual introduction. In itself, the procedure does not ensure that exploration will be durable, but, as we will see, through the links it maintains with the public scene and political power, the effects it produces may be extended in time.

Citizens' Panels and Juries
Consensus conferences are at the center of a galaxy of procedures sharing a family resemblance that are usually grouped together under terms such as *citizens' panels* or *juries*. In every case groups are formed of from twelve to twenty members who are selected so as to be representative of local populations. As in the case of consensus conferences, members are laypersons with diverse educational and social backgrounds. Experts and established interest groups are consulted. At the end of the discussions, which never last more than a few days, the group presents recommendations and proposals. The only real difference is that the consensus conferences deal with problems that arise at a national level, whereas citizens' panels and juries are generally more sensitive to local aspects. Furthermore, they are often asked to make conclusions that are as close as possible to decisions to be made; in some cases, the terms of reference even expect a verdict to be given, the jury or panel having to arrive at concrete recommendations. The results of the work are presented at public meetings and give rise to widely circulated reports.

According to George Horming, more than 100 citizens' juries or panels have been organized since 1970.[15] At present there are no reports or detailed evaluations available. For the most part these consultations have taken place in Denmark, the United Kingdom, and Germany. To give an idea of the variety of procedures adopted, we take the example of citizens' panels organized in Germany, in six towns of Baden-Württemberg at the beginning of 1996. The theme chosen takes the form of a simple question: "How can CO_2 emissions be reduced so as to ensure the conditions of durable development and avoid the dangers of harmful climatic changes, knowing that, since the World Conference at Rio, the German government is committed to reducing its emissions by 25 percent before the deadline of 2005 (the calculation being made on the base of 1987)?" To show the different strategies possible, the groups are presented with three scenarios. One focuses on almost exclusive recourse to the most high-performance technologies, the second on the conservation of resources, and the last on a change of lifestyles. Computer simulation models are devised so that the panels can explore the consequences of different decisions. To form the different juries, 1,000 people are selected in each town on the basis of the electoral roll and, finally, 120 are chosen, or 20 per panel. The same question is posed to each panel: "What strategy should be adopted to obtain the 25 percent reduction?" After training and discussion with experts and interest groups, the panels work out their own scenarios. In all, 53 different scenarios, with commentaries, are presented. Each panel also tries to establish the explicit list of criteria it has used to compare the scenarios. The organizers analyze the set of scenarios in such a way as to draw lessons of general significance. In this way it is noted that all the panels agree on the following points:

• The preferred strategy of energy economy consists in resorting to the most effective techniques.
• It is not deemed realistic to want to reduce CO_2 emissions by basing oneself on a change in behavior.
• There is no agreement on the choice of substitute technologies, nuclear power being the object of pronounced controversy.

Then the organizers assess the measures intended to reduce consumption in terms of the degree of consensus between the groups. It turns out that it would be necessary to resort to measures supported by only 65 percent of the panels to achieve the objective of a 25 percent reduction. If the rule was to adopt only those measures agreed by at least 80 percent of the panels' members, then only a 13 percent reduction would be achieved!

Using the evaluation grid we have proposed, it is easy to verify that this procedure leads to conclusions close to those we came to for consensus conferences. The detail of the forms of organization is slightly modified by the fact that the problems discussed have a strong local component and that there is a stronger imperative to reach a decision. In reality, citizens' juries do not go as far as the consensus conferences in the exploration of dialogic procedures. With regard to the production of knowledge, they are at best limited to establishing possible thematic priorities for secluded research, but without ever encouraging any form of dialogue between specialists and laypersons. With regard to the composition of the collective, the procedure is obviously very limited, since, in the case cited, the general will is posited as a constraint and not as a result. The procedure ensures an open inventory of established positions (by privileging local points of view), but by imposing a framework that is close to that of delegative democracy. (For an example of a procedure that combines a local and a national approach, see box 5.1.)

Entry into the Public Space

Each procedure can be evaluated in terms of its degree of dialogism, that is to say, in terms of its greater or lesser ability to facilitate and organize an intense, open, high-quality public debate. The greater this ability, the more difficulties linked to the double delegation are circumvented. But what kind of procedure would it be that was satisfied just with organizing discussions? What kind of hybrid forum would it be that could be summed up as no more than a space of exchanges? All those who have studied the implementation of the various procedures presented here repeat over and over again that, in the end, the effectiveness of a procedure depends on how well it is integrated in the political decision process.

The worst pitfall to be avoided is that of open, fruitful debates that the decision makers do not take into consideration when they make their decisions. This formulation has the advantage of emphasizing an important question. But things are not so simple. As we will show in the next chapter, the catch-all notion of "decision" should be revised when we are immersed in the dynamic of a hybrid forum. The construction of information to enlighten a confused decision maker does not matter as much as the establishment of the to-and-fro movement between exploration of possible worlds and exploration of the collective. The sole raison d'être of dialogic procedures is the gradual production of a common world.

A procedure could not ensure this kind of dynamic on its own. Let's take the example of numerous consensus conferences that have borne on themes linked to genetics. The expression "biotechnological democracy" has been used by some to give a synthetic description of the limits of the procedure. Levidow maintains that the conference held in France and in the United Kingdom on the theme of GMO have contributed powerfully to focus debates on the notion of risk and more particularly on that of risk directly associated with genes, while passing over in silence not only more specifically political questions, like those around the organization of mar-

Box 5.1

The Conventions (*États généraux*) on health

The *États généraux* on health, organized in France on the initiative of Minister of Health Bernard Kouchner between the summer of 1998 and the spring of 1999, shows how local and national approaches can be reconciled, while illustrating the importance of the conditions of implementation of procedures for the very dynamic of the debates. Apart from the organization of a set of public conferences and opinion polls, they gave rise to a series of regional citizens' conferences, each focusing on a specific theme. The cumbersome general architecture and the short notice (10 months from the initial announcement to the final synthesis) created a chaotic but in some respects ultimately productive dynamic. A group of national experts constituted a dossier for each regional theme. During the training sessions, the jury could reformulate the initial questions and choose experts who would be auditioned during the final public session. Then there was a deliberation session in which the experts answered questions from the jurors, resulting in the formulation of a set of policy recommendations.

These citizens' juries helped to make the diversity of expectations regarding individual and collective health visible. Their recommendations were all the more relevant when certain conditions of deliberation were fulfilled. Significant differences appeared between the regional juries. Depending on the case, there was greater or lesser autonomy with regard to the questions privileged by the professionals: as one might expect, in the case of the forum devoted to rare diseases, the patients' spokespersons were able to make their presence and concerns felt. Some juries, like the one working on aging, succeeded in enriching the initial formulation of the question (extending it to problems of living conditions and retirement). Others remained imprisoned by the framework laid down and were sometimes captured by some professionals, with non-specialists being relegated to the ranks of mere spectators. All in all, one of the most interesting contributions on the procedural level is the decentralization of the forums. This consultation also had a direct political effect by giving the Health Minister proposals and arguments for a law, eventually passed on 4 March 2002, that substantially strengthened patients' individual rights (medical information, access to their personal file) as well as collective rights (representation, participation).

kets, the future of certain professions, or North-South relations, but also ethical questions on the manipulation of the living.[16] Obviously he is not wrong. The subjects not tackled by the two conferences cited are still little explored at present. For pertinent formulations to appear, for arguments and positions to be built up, collective work is needed, which cannot take place within the framework of consensus conferences alone.

We must, however, note that this analysis is partial. This work is underway and is taking place outside of the strongly marked moments of the conferences strictly speaking. This does not disqualify the procedure, but quite the opposite. A consensus conference is a limited intervention, but there is nothing preventing the experiment from being repeated. This is precisely what happened in the United Kingdom with the organization of a first conference in 1994, and of a second in 1998. In 1994 the members of the panel remain prisoners of the rhetoric of risk. In 1998 they broaden the field of the debate, asserting that agriculture must be transformed in order to move away from intensive methods, and in a way that aims for a weak use or no use of chemical or artificial substances, such as pesticides and chemical fertilizers. On this hypothesis, the panel considers transgenic cultures to be pointless, since the enemy they are supposed to combat is destined to disappear. As a result, it radically alters the perspective by now asking itself about the best form of organization of agriculture. This is a precious example. It makes it clear that if the consensus conference, envisaged as a limited, occasional intervention, does not enable us to go beyond the introduction of already constituted points of view into the public space, it may, on condition that it is repeated, become a powerful tool not only for registering the emergence of new identities and demands, but also for getting them taken into consideration in the public debate.

A hybrid forum has a dynamic. Every consultation implemented by this or that procedure brings about the emergence of groups and opens up new lines of collaborative research. Experimentations are organized and lessons are drawn, which opens up the field of possibilities. New consultations are launched resorting to the same or different procedures. The search for a common world continues. This continuity is possible only if an infrastructure exists ensuring that exchanges and contacts are not broken off between the moments of consultation and discussion organized by the procedures. It is one thing to define who will participate in a collective negotiation, and in what mode, but it is quite another to provide for the table around which the protagonists will sit, the room in which the table will be installed, the building that will accommodate it, the roads of access, and so on. Without such an infrastructure, the best procedure in the world is just a waste of time. You consult, debate, sometimes you even decide, and ... whatever will be will be!

But a durable framework is required so that whatever will be will be, that is to say, so that collective exploration and learning continues. Having met and then parted, the actors must not completely disappear. They must be able to continue the dialogue with new partners when they judge it to be useful. This stable and open framework is what we call the public space; a public space which can function only if it is fitted out, organized, structured, and shaped to allow the implementation of dialogic procedures.

In what must such an organization consist? Although some elements emerge on the basis of available experience, we are still far from being able to give satisfactory specifications. It would seem that associations, the media, and the public authorities must all contribute if the dynamic of exploration of a common world is to be maintained. Associations give emergent groups their first existence and recognition as well as their first means of expression. The media provide an infrastructure that gives publicity to positions and controversies, to the structuring of which they obviously make a major contribution. Public authorities keep the dialogic procedures in good working order at the same time as they act as a source of support and establish structures of coordination; they allocate resources so that collaborative research develops and the costs of the composition of the collective are taken care of.

Two contrasting examples will enable us to suggest the crucial role of these three elements and to show the variety of dynamic configurations into which they enter or that they may give rise to.

Let us consider first the case of the AIDS epidemic and its management, the public dimension of which has been very carefully analyzed by Nicolas Dodier and Janine Barbot.[17] Confining ourselves to France, from the mid 1980s until the end of the 1990s patients' groups or associations are formed to make the voices of those infected by the virus heard. These voices very quickly reveal themselves to be multiple, as also are the modes of action they propose. *Aides, Arcat-Sida, Act-up, Actions Traitements, Positifs*: if these associations are distinguished from each other by different forms of involvement, all of them, in their own ways, are interested in research, and notably in drug trials. They intervene directly in the debates on their modes of organization, on the methods to be used, and on the indicators to accept when deciding on the effectiveness of the different molecules tested. This collective involvement results in the establishment of what turns out to be a quite original organization of trials. It reserves an important role for the patients, who prove to be impatient and active, and the expression of their point of view, which carries great weight both when defining the rule adopted and when it is a matter of halting a trial or drawing conclusions from it. These associations do not speak with a single voice, and each of them finds an echo in the positions it defends with this or that group of doctors, researchers, or political decision makers, or even in the world of the pharmaceutical industry or of public research. Under the pressure of these associations, the range of forms of organization of research, methods employed, and paths explored opens up. *Actions Traitements* does not refrain from negotiating compassionate protocols with industrial secluded research, without this challenging the monopoly of the enterprises on this research, while *Act-up* and *Aides* are quick to plunge into scientific matters to discuss the meaning of the increase of viral loads as indicators of the effectiveness of the molecules tested. In addition, these two associations put a lot of effort into informing the public about medical matters, while demanding a halt to experimentations on double therapies and their replacement with triple therapies, and this without full adherence to the double-blind procedure. The clear and sharp division between laypersons and specialists gives way to a multiplicity of coalitions and configurations that make patients, clinicians, biologists, pharmaceutical laboratories, and public research bodies collaborate in sometimes competitive and sometimes complementary programs of research and experimentation.

At the same time as the forms of organization of collaborative research diversify, the public space opens up to take in new identities that take shape with these co-operations and develop along with them. Some associations, such as *Act-up*, radically assert the patients' social and political identity, resorting to extreme and spectacular modes of action[18]; others are more inclined to introduce the epidemic and the concerned groups into existing frameworks of management. Some insist on the fact that the patients are above all victims and struggle to get this identity recognized; others strive to play the role of mediator between patients and institutions. The clear-cut distinction separating patients on one side, and medical and scientific institutions on the other, is replaced by a continuous range of different forms of collaboration. In this way, within this multiple and crowded movement, strong assertions of identity which are not interested in the collective coexist with other, no less strong claims in favor of full integration in a recomposed collective. The first attitude leads to an original innovation: that of the category of *political seropositive*, which permits militants who are not seropositive from a biological point of view to take on the stigmata for the purpose of de-stigmatization. "We are all seropositive" is the slogan adopted by these militants who, by denying the disease in itself, seek above all to impose a particular, singular identity that there is no question of relinquishing in the face of any kind of general will. The second attitude puts to the fore the affirmation of rights possessed by every human being, by every citizen, whether or not they are affected by the virus. In the name of a community in which everyone has the right to the same treatment, identity is asserted on the basis of the affirmation of the (potential or already broken out) disease and rejection of the inequalities it brings about.

What an amazing HIV forum! The associations defend very different positions and orientations. Some defend their identity as a priority, refusing to address the question of the collective, while others develop an acute sense of concern for the common good. Some have no qualms about supporting secluded research, while others struggle for intense cooperation between secluded research and research in the wild. And, rather than tear each other apart and confront each other, they play on their complementarities and even develop common projects. It is because the associations are present on every front and because they occupy the whole of the space of dialogic democracy that the latter succeeds in asserting itself. Radicalism and a propensity to negotiate mutually support each other. *Aides* can lean on the strong claims of *Act-up*, which in turn can adopt a hard-line radicalism without risk of a break. How does collective exploration, which is quite clearly facilitated by the fragmentation and diversity of the associations, prevail over the centrifugal forces and break-up? In other words, how is a (relatively) unified space constituted that makes possible the deployment of a dynamic of exchanges and the actions to which these give rise? The history and organization of the French HIV forum make it an ideal case for bringing out elements of an answer.

Confining ourselves to France, the creation by the public authorities of the CNS, the AFLS, and the ANRS[19] played a major role in structuring and unifying the hybrid forum or forums dealing with AIDS. Since its creation, the ANRS, to which the state delegated the organization and coordination of research, has had a considerable budget for supporting researchers. It became the veritable project manager of research programs, mobilizing pharmaceutical laboratories, public research bodies, clinicians, and representatives of the patients. The ANRS was very quickly transformed into a central authority which became all the more inescapable since it was responsible for following the clinical trials that turned out to be at the center of the debates and controversies. To coordinate their actions within the ANRS and avoid diluting their presence in the midst of all the other protagonists, various associations decided to create a group called TRT5 in which they confronted and discussed their positions in order to arrive at a common point of view on the main issues. The creation of an autonomous body to which the public authorities delegate some of their jurisdiction thus leads the associations to come together with each other, but also with other parties, and to do this on subjects that in some cases directly affect the organization of the production of knowledge.

Media coverage of the demands and controversies is the second element of explanation for the strong structuring of the hybrid forum. Every actor who intends to enter the public space organized by the network of different media is constrained by the connections and confrontations this network sets up to argue and justify their point of view, as well as possibly to consider the points of view developed by other actors also present in this space. In the case of AIDS, the merit of the first associations in bringing the debate into the media universe should be acknowledged. They were helped in this, moreover, by some of the scientists themselves and by political decisions makers (think of the long controversy between Montagnier and Gallo on French precedence in the identification of the virus, and then the summit negotiations between Chirac and Reagan to decide on the patent rights for diagnostic tests). Once they appear in the public media space, the different protagonists (associations, clinicians, biologists, and others) found that they were forced to take account of each other's respective positions, even when they disagreed with each other. The media—the press, radio, television— made a powerful contribution to making possible, organizing, and structuring the debate. According to their own specific logics, they established relations between the actors and their positions and made them mutually perceptible. In this way they took part in the construction of the infrastructure necessary for the double exploration and the to-and-fro movement it presupposes.

Finally, let us observe that the keystone of this space is HIV itself. In fact, patients' associations, as well as pharmaceutical laboratories, clinicians, or biologists, have this composite, hybrid being in common: the "patient affected by HIV." The virus circulates in different forms in the patient's veins, in the researchers' tubes and trials, in scientific publications, and in the dossiers assembled by pharmaceutical laboratories to get authorization for drug trials. This same circulating virus links these actors and imposes a common destiny on them beyond their oppositions and differences. Here again, we should emphasize the voluntary and effective role of associations in organizing contacts with scientific and industrial communities, and in making effective and legitimate this "impure" science, with the aim of establishing a link between "their" virus and the one studied by laboratories and firms.

By being constrained not only to position themselves in a public space strongly structured by the media, but also to coordinate themselves in order to access resources and legitimize their action, the different actors participate in the gradual construction of a unified and centralized hybrid forum. The conditions for a possible and necessary movement to and fro between scientific exploration and exploration of the collective are created in the same process. Every new result obtained in the laboratory or emerging from a drug trial is taken up again by the associations who evaluate it and discuss its significance and meaning with other associations, as well as with the public authorities, pharmaceutical enterprises, or medical institutions. In the heat of the discussion, which enables the associations to posit themselves by confronting each other, the identities of the different protagonists evolve, their expectations are transformed, and their projects become clearer while being adjusted to each other. In turn, these debates lead the associations to struggle for certain forms of the organization of investigations.

One example, taken from many, will be enough to show the dynamic of this double exploration. When the first trials of triple therapy seem to show their effectiveness, the question arises of whether the standard procedures for testing new treatments should be maintained. Faced with this question, the associations develop different arguments. Some reckon that the first indications should lead to the abandonment of the previous experiments and that the rules in force should no longer be followed blindly. This position leads to an alliance with some pharmaceutical laboratories, which may have an interest in relaxing the protocols to be observed so as to obtain speedier authorization to market their product. Others, on the other hand, argue for the retention of strict rules, ensuring what seems to them to be greater objectivity and rigor in the results and interpretations. Through this episode a dynamic emerges in which the patients' associations are directly involved and which enables both collective identities to be deepened and forms of investigation to be transformed.

This example makes clear what we should understand by the integration of procedures in the political decision-making process. A conventional vision would have lead to the adoption of a procedure of consultation, in this case the Conseil national du Sida (CNS), and to make it a consultative organ of political power. No doubt such a configuration would have resulted in some interesting proposals, but which at worst would have remained isolated, and at best sporadic. It would not have allowed the dynamic we have just described. The development of this dynamic required a triple conjunction of a diversified and free movement of associations, media coverage of the issues at stake, and the government's creation of an agency alongside the CNS responsible for programming collaborative research. The virtuous circle can then unfurl. The diversity of positions on the two axes defining dialogic democracy is made visible in the public space, which makes confrontation between them possible, indeed necessary. The coordination and composition of these positions are also stimulated, but now through the creation of structures devised to organize the confrontation and discussion of points of view, with the aim of arriving at measures concerning the conduct and organization of research as well as the modes of implementation of intermediate results obtained. Of course, such integration lasts only a while. The hybrid HIV forum reconfigures itself. Other forms arise, engendered in part by those that disappear, like that which debates inequalities between North and South in access to available therapies, or like those in which the more general question of patients' rights is raised.

The still recent history of the HIV forum shows that the search for a common world takes some time when there is considerable political and scientific uncertainty. It is a long, slow process in the course of which identities, which are nourished by and in turn give rise to scientific investigations, are worked out, stabilized, and adjusted to each other. This history also shows that this process can maintain and establish itself only if the conditions needed for the construction of durable public spaces are brought together. In the case of HIV, this meant the emergence of diversified associations, the good will and relevant intervention of the public authorities, as well as mobilization of the media, each force leaning on the other two.

The history of the forum of neuromuscular diseases confirms, by contrast, that these three forces are needed to maintain the dynamic of exploration of a common world. In fact this forum followed a different trajectory. A hard struggle was required to achieve integration of those suffering from myopathy into the public decision-making space. Unlike the AIDS forum, which quickly benefited from the initiatives of the public authorities, the neuromuscular diseases forum had only the media to rely on for getting its demands for recognition into the public space. At the start of the 1980s, the AFM was the only association speaking in the name of those suffering from myopathy. Until recently the public authorities remained silent and gave hardly any support to a small association that represents only a handful of abandoned patients. Faced with this passivity, the AFM had only its own forces to count on. To get itself known and recognized, to enter the public space, it chose to occupy the more easily accessed terrains of research and the media. The telethon[20] enabled it to make a double coup, that is to say, on the one hand, to bring together money to finance laboratories and require them to disclose their results, and, on the other, to invite itself into French homes for a week-end. The association thus committed itself to what it calls the genetic road, in the process building up French genomic research potential.

The AFM thus found itself doubly isolated: first in being condemned by the public authorities to limit its action and discourse to scientific exploration alone, second in taking initiatives on its own in the conduct and organization of this exploration. This double isolation led it to devise exceptionally original forms of cooperation between secluded research and research in the wild. But it denied it the possibility of integration in a larger public space in which its identity could have been recognized and discussed and in which it could have taken part in the recomposition of the collective. It has constantly had to demonstrate its strength in order to compensate for the weakness of its interlocutors. Not until much later was the association, having acquired a recognized place in the media and public research through hard struggle, taken into consideration by the state, which accepted to listen to and support it in its battle for citizenship. The creation of the Alliance, a grouping that brings together several dozen associations of patients suffering from rare diseases, continues this trajectory. It is established in partnership with the public authorities on a European as well as a national level. Its recognized weight enables it to be active in research programs on new medicines, in the organization of the medico-social environment, and in the struggle for recognition of the rights of handicapped persons.

The comparison between the history of the AIDS forum and that of the forum of neuromuscular diseases gives us a better understanding of what is in play when we invoke the necessary integration of procedures in the political process. It is not so much a question of ensuring the link with a hypothetical decision making, as of enabling different groups, and notably those that are emerging, to enter and move about in the public space. This is the price for the dynamic of the double exploration to be able to unfold, orchestrated by dialogic procedures and their implementation. But the latter are powerless if the actors lack this space. In fact, organizing debates and launching collaborative research is of no use if there is no mechanism enabling actors to leave their private sphere, if there is no accessible space of evolving confrontation.

The two adventures of AIDS and neuromuscular diseases contain another lesson. The production of this space presupposes the combination of three forces. The first enables groups to be formed and associations to develop, exist, and maintain themselves ; this requires favorable legislation protecting associations from any state or economic takeover. The second provides groups with the means to see and make sense of each other, each group finding that they are affected by the others; the media obviously contribute to this enterprise. The third is the state; it not only makes the deployment of the first two forces possible, but in addition enables groups to discuss whom and what the collective should comprise. The political realism of procedures is decided by the combination of these three forces. As for the trajectories taken by the exploration, these differ from one forum to another. Some rapidly close up, others endure, are transformed, get going again, split, or fuse. Every forum is a singular history.

   Dialogic democracy, like delegative democracy, is a matter of procedures. These are devised to promote the organization of a debate that is respectful of scientific and political uncertainties so that it is better able to take responsibility for them and manage them. The inventory that we have proposed, albeit partial[21], highlights the road that still must be traveled in designing dialogic procedures. None of these procedures has the required profile to allow for the effective deployment of the two explorations and the constant coming-and-going between each of them implicit in the dynamics of dialogic democracy. In fact it seems unlikely that only one procedure would suffice. The temptation would be to favor and to institutionalize a single procedure (for example consensus conferences) and then to consider that the question of dialogic democracy is solved. But dialogic democracy implies tools allowing for constant reflection and debates on procedures (should one use available ones or design new ones?), their implementation and their evolution. It is more probable that different procedures would need to be combined, depending on the state of the controversies and their degree of maturity. The social sciences surely have their part to play in collective reflection on the "right" procedures. The proposed grid illustrates the approach consisting in starting with existing experiences, and then enriching and reviving them. The case of the AIDS hybrid forum shows, for instance, that (at least in the French case) the procedural device has to go relatively far in establishing organizations which equip research with a collaborative dimension. These organizations are still rare and little is known about them. Many experiments need to be undertaken to identify their characteristics, the conditions of their implementation, and their evolution. The social sciences could certainly contribute meaningfully when it comes to setting up such experiments and drawing conclusions from them.

Finally, setting up a dynamic of exploration to which the debate contributes, and by which it is fed in turn, is possible only if a structured space of communication and perception exists that allows for and facilitates the to-and-fro movement between scientific investigation and the adjustment of identities with a view to composing the collective. This is where identities are affirmed and discussed, where experiments take place and lessons are drawn, where the multitude of decisions are woven that re-launch investigations or adopt *modus vivendi* which are deemed to be acceptable. No procedure by itself can guarantee the existence and durability of such a space, which is born from the conjoined action of the state, the media, and the groups themselves. That is why the notion of hybrid forum is not reducible to that of procedures.

The spatial metaphor certainly has limits. It nonetheless has the immense advantage of making another extremely dangerous metaphor pointless: the linear metaphor. No, the debate does not prepare clear-cut, definitive decisions which install a before and after and which permit us to be rid of the past once and for all. It is made up of a multitude of micro-decisions, each of which, taken individually, is not irreversible and which nonetheless ends up by forging a robust network, as the debate matures and learning advances. To talk of space rather than of linear and sequential processes is ultimately to prepare oneself for doing without the burdensome notion of decision. You do not decide in organized hybrid forums. You take measures. You take measures in the metrological sense, in order to draw up a map of overflows, of concerned groups, of their positions and of their relations. No debate is possible without this cartography. You also take measures in a political sense, in order to maintain and re-launch the dynamic of dialogic democracy. This does not mean inaction but what it is better to call *measured action*. We will now turn our interest to this. On the way we will encounter the discourses and reflections on the now-famous but also always-debated principle of precaution.

6 Measured Action, or How to Decide without Making a Definitive Decision

The dialogic democracy devised by hybrid forums rests on procedures of consultation that do not sit easily with the idea of sharp, clear-cut decisions. The search for a common world presupposes in fact careful consideration of backward steps, that actors avail themselves of the means to be able at any moment to return to abandoned options, and that evaluations are constantly revised in terms of new knowledge and points of view. This constant attention is far from being synonymous with indecision and temporization; it defines what we propose to call *measured action*.

Actors immersed in hybrid forums and engaged in socio-technical controversies have themselves carried out this work of revision. Faced with situations of great uncertainty, but without this leading them to think of renouncing action, they have revised the notion of decision, inventing notably the now-famous precautionary principle.

Because we have chosen in this book to follow the actors in their work on the elaboration and implementation of new procedures together with new conceptions of political processes, at this point we are naturally interested in the precautionary principle. This notion is the still-unfinished result of an intense activity of research in the wild in which professional legal experts have been quick to take part. This explains the profusion of definitions, but also their instability, even though these divergences are starting to decrease. Our objective in this chapter is to do justice to this richness while striving to show the relevance of the notion of measured action in accounting for this turmoil.

In plain language, the idea of precaution may be formulated as follows: "When there is doubt on the existence and impact of potentially negative effects, as well as on the identity of groups concerned by these effects, above all do not refrain from action. Take steps instead to evaluate the danger and search for the means to control it." This notion is perfectly in line with the aim of this book for two main reasons. First, precaution designates an active, open, contingent, and revisable approach. It is exactly the opposite of a clear-cut definitive decision. And then, this approach rests on a deepening of knowledge—but not only of the knowledge provided by the scientific disciplines of secluded research. Proportionality of actions, social acceptability, and economic cost also have their place in deliberation. Precaution is therefore a positive approach of assessment and management in situations of great uncertainty. While it is of considerable theoretical importance, the actual conditions of its implementation with regard to the environment and health raise a great many questions.

This is all the more so as precaution is the victim of its success. In fact, like many new notions, the term spreads everywhere and it is given the most extreme interpretations. This keen interest reinforces the confusion and does harm to the idea more than it helps to clarify and consolidate it. The idea of precaution is often emptied of its specificity and considered as a synonym of prevention, of focused attention, or any security arrangements. Even the scientific press sometimes adopts such assimilations. In a recent editorial in the journal of scientific information *La Recherche*, titled "Human errors and the precautionary principle,"[1] the author proposes to extend the precautionary principle, as formulated for questions to do with the environment, "to the field of collective errors, entailing risk of grave and irreversible collective damage, on the evaluation of which, at a given moment, science feels helpless." It is a generous idea! But the examples given in support of this formulation reveal a double confusion. The first confusion leads the author to overlook the fundamental difference between the prevention of known risks and precaution concerning situations of uncertainty. The second leads him to identify the failure of a system of prevention with a defect of precaution. "A serious rail accident in Great Britain due to failure to stop at a red light, a nuclear accident at Tokaimura following a technical mistake, an error at NASA in the calculation of the trajectory of a satellite heading for Mars, hepatitis infection through transfusion. Every time, at a given moment precautions have not been taken to reduce the risk of catastrophe," he asserts. Now, the four cases cited involve, on the contrary, domains in which considerable experience has been accumulated and in which safety procedures have been increased. Science is not at all "helpless" when faced with these situations, but the existence of multiple procedures limiting risk in no way guarantees the absence of individual or sequential human errors.

Apart from these terminological confusions, which still attest to the fuzziness of the concept in everyday language, precaution also has its Cassandras. These prophets fear that it will lead to "irrational," economically costly decisions that immobilize research. They rely on maximalist, dramatizing definitions in order to stress the potential dangers of the principle in an attempt to guard against new requirements. The scientific director of a pharmaceutical group put it this way: "Today, by virtue of the precautionary principle, any activity in which theoretically there could be a risk should not be undertaken unless its outcome can be predicted with precision.... The principle becomes a real encouragement of paralysis."[2] Some economists, anxious to reveal the excessive character of some formulations of the principle, endeavor to criticize what they see as bad uses of precaution and the danger this presents for the collective management of risk. Studies on the history of the precautionary principle[3] show that in the 1990s emphasis was often placed on the lack of legal precision of the concept, and on the dangers that its use could entail for the freedom of research. Others limited their definition of the principle to a strengthening of the state's responsibility as a guarantor of the people's safety. In this way they sought to provide clearance for the economic actors who are the primary generators of situations of danger by reducing precaution to a supplementary provision of administrative police. The most virulent critics of precaution are often from the medical world. Some health professionals think that precaution contributes nothing new to what has long been the practice with pharmaceutical products and in epidemiology. Others, faced with a series of proceedings for administrative and then penal responsibility, instituted in the wake of the dramas of HIV-infected blood, worry about hepatitis C or nosocomial infections in the hospital milieu. A head doctor of a hospital department writes: "Some kinds of behavior must change: we must struggle against the precautionary principle, which aims very high without real validation but in the name of which costs increase. This of course greatly benefits the pharmaceutical industry which knows how to utilize it, like some doctors moreover, less concerned with doing the best they can than with avoiding legal proceedings."[4] For all these skeptics or Cassandras, precaution therefore tends toward an irrational, costly, and counterproductive approach.

In order to extract the terrain from a set of received ideas and prejudices, we will begin by presenting what precaution is not, then we will show the notion's characteristic zones of uncertainty, and finally we will develop the main approaches to which it has given rise and in which it begins to prove its worth. In this way we will gradually draw out the notion that gives unity to this approach: that of measured action.

What Precaution Is Not

A precautionary approach must not be confused with the prevention of risk. It is not an incitement to abstention, it does not require the demonstration of zero risk, it is not an obstacle to scientific and technological research, and it does not result in a supplementary penalization of the decision maker's responsibility. Is this list clear enough? Not as clear as we might suppose. To see the difficulty of clarifying these misunderstandings, see the statements in box 6.1 and attempt to check off those that seem to

Box 6.1
Find the errors

1. "The continued import into Great Britain of animal feed after the identification of its contamination by prions in March 1996 was contrary to an elementary principle of precaution."
2. "With regard to protection from health risks, it is no longer a matter of acting once the damages have been observed, but rather of evaluating them *a priori* in order to act. We still need to know how to grade a response that is in proportion to an uncertain danger."
3. "Suspension of authorization for the cultivation of genetically modified corn is in line with the precautionary principle which requires a decision maker to embark on a policy only if he is certain that it contains absolutely no environmental or health risk."
4. "This new concept is defined by the public or private decision maker's obligation to compel or refuse action in terms of possible risk. In this sense, it is not sufficient for him to model his action on the consideration of known risks. In addition, he must give proof of the absence of risk in light of the current state of science."
5. "The fact of having to act according to the state of technical and social scientific knowledge established at the time of the decision is sufficient to remove responsibility from the person whose activity will later be revealed to have caused damage."
6. "Uncertainty with regard to risks of climate change due to the greenhouse effect does not absolve from responsibility, rather it reinforces responsibility by creating a duty of prudence."
7. "The precautionary principle is an approach to the management of risks that is exercised in a situation of scientific uncertainty, expressing an exigency of action in the face of a potentially serious risk without waiting for the definitive results of scientific research."

you to correspond to precaution. Did you hesitate for a moment before some definitions that left you perplexed? Have you recognized, without doubt crossing your mind, that only formulations 2 (*a priori* evaluation of risks and graded response), 6 (precaution reinforces the decision maker's responsibility), and 7 (definition given by the DG XXIV of the European Commission) correspond to the precautionary principle? No doubt, but only after a moment of reflection. That is why some words of commentary could prove useful.

Precaution and Prevention

The first confusion to be avoided is that between precaution and prevention, between the management of uncertainty and management of an identified risk. This is the most common error. The continued importing of animal feed after identification of contamination by prions in March 1996 (formulation 1) was contrary to a preventive approach and not to the precautionary principle. It was in fact at this date that the hypothesis of the possibility of the transmission of BSE from the animal to humans was recognized as established. The risk being identified, the public authorities were no longer in a state of uncertainty and had all the information necessary to act in a preventive fashion. On the other hand, there would have been a precautionary approach if measures had been taken to control animal feed and limit human consumption of beef from the start of the epizootic observed in Great Britain from November 1986.[5] The English authorities did not embark on this approach. They waited for scientific proof of the origin of the epizootic before taking the first measures of prevention, which were decided in July 1988. And not until 1992 was the hypothesis of infection by feed made from animals epidemiologically validated. Thus, six years were needed to identify precisely the cause of the epizootic, and ten years to validate the possibility of the transmission of prions from animals to humans. From March 1996, the risk of transmission of "Mad Cow Disease" had thus become a risk which was doubly established: from 1988 for the epizootic, and from 1996 for human infection. The term *known risk* (*risque avéré*) indicates that a harmful situation and its causes have been identified, either through observation (technological or natural catastrophe, degradation of a milieu, clinical records, an epidemic situation) or through probabilistic modeling (correlation between high speeds and the gravity of road accidents, between smoking and lung cancer, or between operational accidents in a nuclear plant). On the other hand, before 1996 the risk of transmission of BSE was potential, as are the risks linked to climate change today. Negative effects, on man or nature, have been identified, but there is as yet no completely validated system of causal explanation. We are still in a context of uncertainty.

The example of the gradual supervision of asbestos also provides a good example of the distinction between the period of precaution and the period of prevention. The dangers of this substance for the health of workers handling it directly were known from the beginning of the 1930s.[6] Beginning in 1975, the state of knowledge made it possible to define the risks due to exposure above a certain level, although controversy continued over exposure to low doses. Measures completely prohibiting asbestos were taken in different industrial countries in the 1970s. In France, such measures were not taken until 1996.[7] The risk of pulmonary diseases was sufficiently known from 1975 on for real preventive measures to be taken, the most radical being prohibition. Before 1975, the measures that could have been taken would have come under precaution in the face of identified but poorly defined dangers. In France, in July 1975, we were content to prohibit people under the age of 18 from working on the carding, spinning, and weaving of asbestos.

The same type of retrospective analysis can be made on the basis of the case of HIV-infected blood.[8] The first cases of AIDS were recorded in 1980 in the United States and in 1981 in Europe, and the hypothesis of an unidentified causal agent transmissible through blood was first formulated in April 1982. The risk factors became clearer on the basis of clinical observations in 1983. That is why the measures for selecting blood donors taken in France (June 1983) and Britain (September 1983) actually come under a precautionary approach.[9] In actual fact, the nature of the causal agent was first elucidated in April 1984, and the screening test for blood donations became available at the beginning of 1985. The 1983 measure could have been more complete (limits on transfusions, systematic autologous transfusions), but they had some effect. After 1985, all the measures taken come under the heading of prevention and no longer under precaution.

The foregoing examples enable us to see how the gradual transition from precaution to prevention takes place. A potential risk is constructed on the basis of a bundle of indications and hypotheses that are not yet scientifically validated but which permit a warning to be given. Its identification depends on a relationship being established between heterogeneous bits of information, produced by both secluded research and research in the wild, which gradually makes it possible to delimit the uncertainty. Experts and laypersons make use of complementary kinds of knowledge that make it possible to advance in the identification of the danger and in ways of defining it and organizing a precautionary approach limiting possible damage. Once the risk is established (that is, when it is known in its manifestations and explained), subsequent decisions fall under prevention. This does not mean that all uncertainty has vanished and that proof has been provided. But the questions are considerably reduced, and the effectiveness of the measures is put to the test. At the time of writing, the case of BSE provides a good illustration of this change. Certainly, two uncertainties remain. First of all, there is only epidemiological proof for the infection of cattle via the use of animal based feed. That is to say, the correlation between two factors has been demonstrated, but the spread of prions through oral ingestion has not been proved experimentally. And then, the link between the consumption of beef and the development of Creutzfeldt-Jakob disease in humans also rests on an epidemiological hypothesis and not on clinical demonstration. However, the probabilities of the correlation in both cases are such that uncertainty seems reduced and that present measures essentially fall under prevention.

Precaution and Abstention
The second major confusion consists in identifying precaution with a mandatory rule of abstention (as claimed in assertion 3), which would remain in force as long as certainty of safety was not established (like the definition of a level of non-harmfulness of a chemical product, or the availability of an absolutely trustworthy screening test). Some critics of the precautionary approach reckon that this reference makes all action impossible so long as there is no absolute proof of the absence of all danger. This interpretation is contradictory. It refers to the idea of mandatory abstention from action while envisaging demonstration of a product's non-harmfulness, which at the very least is an action. It would be more coherent to speak of suspension of the execution of an action (marketing of a product, construction of an installation) and not of abstention from action, since the actor interested in acting is on the contrary encouraged to provide proof of harmlessness.[10]

This conception of precaution is an abusive interpretation promoted by decision makers who would like to benefit from a limitation of their responsibility. But such an absolutist incitement to abstention does not appear in any legal text of general significance. In practice, only some radical ecologists have given precaution this meaning. Thus, the famous Greenpeace declaration is often cited, in which it is asserted that "there must be no discharge of waste into the sea until the harmlessness of this waste has been formally proved." The context of this declaration relativizes its significance. For the ecologists it was a matter of being assured that the necessary authorizations for carrying out such discharges demanded a maximum guarantee on the choice of burial sites, on the volumes disposed of, and on the resistance of containers to marine corrosion. Moreover, with this warning, Greenpeace was only recalling the terms of the declaration of the international conference on the protection of the North Sea in 1987. This provided in fact for a control on emissions of "the most dangerous substances," even if a link from cause to effect had not been established between these substances and harm. It being a matter of toxic industrial waste, we may reasonably think that to a great extent the risks were known for their effects on the continent, although they had not yet been observed in a marine milieu. In short, there was as much prevention as precaution in this attitude. If there is precaution, paradoxically it is found in the official formulation rather than in the Greenpeace declaration: "To protect the North Sea from the effects of the most dangerous substances likely to be harmful, a precautionary approach is necessary, which may require that measures are taken to limit the deposits of these substances, even before a relation of cause and effect has been established by incontestable scientific proofs." The objective was to protect the ecosystem by limiting emissions at source by various methods (reduction of quantities, employment of better technologies). If precaution had been made equivalent to a principle of abstention, a prohibition of any discharge would have been formulated, inasmuch as the absence of harmfulness to the milieu could not have been established scientifically. There is no question of this. The declaration is conditional and envisages instead various measures of reduction without laying down any principle of the prohibition of discharges.

On the theoretical level, some authors attribute this identification of precaution with abstention to German legal and philosophical thought. However, it would be a bit premature to claim that. In fact, in the mid 1970s the idea of precaution was formalized in German, first legally (the law on chemical products and its extension to the environment in 1976)[11] and then philosophically (notably in 1979, when Hans Jonas published *The Imperative of Responsibility*[12]). In his 1979 work Jonas puts forward what, at a conference in 1957, he called the "ethics of responsibility," which in reality is a theory of action. Jonas claims that today "domination takes the place of contemplation." He sees knowledge and the possibilities of the transformation, indeed destruction of the world as now being intrinsically linked. From this Jonas deduces the need to develop principles and practices that lead human beings to self-limitation. Referring to the Nazi holocaust and various environmental catastrophes, he develops what he calls a "heuristics of fear," which should not be understood negatively as a call to scientific and technical immobility, but as having to lead to the "anticipation of the threat itself." Nature has long been "the subject of history dictating its own laws to man." Now its status has changed. Inasmuch as it has become an object of domination, it must also become an "object of responsibility." Faced with the will to power, with the irreversibility and unpredictability of the processes unleashed by human action, like Hannah Arendt, to whom he is very close, Jonas proposes a constant reference to moderation, respect for limits, and responsibility. For this reason he is also concerned with the concrete translation of his analyses into practices: "In contemporary ethical reflection, there is quite a lot of padding in good intentions and irreproachable motives, which affirm that we can take the side the angels and that we are against sin, that we are in favor of prosperity and against decline. We need to attempt something more solid here." Jonas calls for a break with technical ideology, and the meaning he gives to the principle of responsibility is meant to be a guarantee of future life. Everything that may undermine humanity in the long term should be avoided. However, it does not seem to us that Jonas takes the step of making the "principle of responsibility" a "principle of abstention." Certainly, he envisages radical, indeed authoritarian ways: "What is clear in any case is that only a maximum of politically imposed discipline is able to realize the subordination of present advantage to long term command of the future."[13] But further on he appeals to a "soft" and "enlightened" tyranny, and in a section entitled "Progress with precaution" he says that, faced with contemporary scientific uncertainties, "while waiting for the certainties resulting from [scientific] projections to become available—especially in view of the irreversibility of some of the processes unleashed—prudence is the better part of valor and is in any case an imperative of responsibility."[14] Uncertainty is presented as humanity's new destiny. Moral consequences result from this that the activities of precaution help to translate into practice.

Precaution and "Worst-Case Scenarios"

Those who wrongly identify precaution with a rule of abstention also reckon that by referring to "worst-case scenarios" this approach removes all rationality from decision making. Precaution would lead to a reasoning based on holding the most extreme hypotheses to be probable—for example, that the epidemic of prions transmitted through food will spread to every animal species, or that the cultivation of GMO will create irreversible mutations everywhere and will strengthen human resistance to all antibiotics, or that global warming due to increasing CO_2 emissions will generate irreversible catastrophes on every continent. No organized reaction would be in a position to respond to such scenarios, which would no doubt result in every decision being blocked.

Here again the formulations are extreme. They are usually found in the writings of journalists who like to play on the emotions of the public, and in the discourses of industrial interest groups opposed to the regulation of chemical products.[15] They do not appear in any reference text. Reference to "worst-case scenarios" corrupts the serious notion of "the most pessimistic hypothesis" which is generally used in classical procedures of risk assessment. To evoke the "worst" hypothesis is not to say that the worst will certainly come about! Faced with uncertainty, it is a constructive reminder that precaution encourages the consideration of all hypotheses, even the most marginal. Hence the importance of "whistleblowers" and other "birds of ill omen" who draw attention to facts which are isolated and enigmatic but likely to announce broader attacks. This was the case of the English veterinarians who in November 1986 anticipated the BSE epizootic. Maybe this is how we should understand the formula of the government commissioner of French Supreme Court who, in one of the compensation proceedings in the affair of infected blood, stated: "In a situation of risk, a hypothesis that has not been invalidated should be held provisionally as valid, even if it has not been formally demonstrated." If reasoning based on worst-case scenarios can only lead to an impasse in decision making, on the other hand, the absence of a formal demonstration of the existence of dangers should not be a pretext for exemption from various forms of mobilization.

Precaution and "Zero Risk"
Another way of raising the same specter consists in presenting the precautionary approach as one that has to guarantee a situation of "zero risk." In precaution some authors want to see only a synonym of absolute safety, action which would lead more to "sealing off" than to an exploration, to the pursuit of complete security at any cost. This confusion is very widespread. At a session of the French Academy of Sciences devoted to the precautionary principle (January 2000), one of the speakers gave great emphasis to "the illusion of zero risk" that this principle would introduce.[16] Similarly, in an editorial of the medical association *Bulletin* of December 1999, the president writes:

The precautionary principle has been extended to the health domain where in view of scientific progress risks are increasing with technical development. Controlling technology has become an obsession and the doctor will be driven to justify his medical actions by providing proof that he has envisaged every risk and taken all the necessary precautions.... If this pursuit of zero risk is unanimously shared, it must be firmly stated that it cannot be applied either to medical practice or to research....[17]

This fantasy is broadly shared, as is testified by another recent document produced by experts of the State Planning Commission when they judge that "the public authorities are condemned to a transitional phase of excessive precaution in order to avoid conflicts with public opinion," which would put the decision maker in a situation in which he is "called on to foresee everything, including the unforeseeable."[18] These mistaken and dramatized versions of the precautionary principle are evidence of the frequent confusion between the obligation of means and the obligation of results. In theory this distinction is very clear in health matters, since the doctor is only ever held to the first. However, an inadequate formulation of the French Supreme Court has sown confusion and unfortunately provided arguments for the Cassandras of precaution. In the Supreme Court's "reflections on the right of health" it is noted that "it is not enough for [the private or public decision maker] to model his conduct on the consideration of known risks. In addition, he must provide proof of an absence of risk."[19] The duty of precaution is thus understood as going beyond the duty of prudence and diligence that characterizes the obligation of means. But does it thereby become an obligation of result? Apparently it does not, for a few pages later the Supreme Court dismisses the usefulness of precaution in the health domain; it thinks that the meaning given today to the notion of created risk is sufficient to cover demands for compensation for victims of medical accidents. Nevertheless, this analysis, also found in other countries, is very unsatisfactory owing to its ambiguities. It refuses to envisage precaution as a model for the management of emergent dangers, one which encourages the deployment of appropriate procedures of investigation. And, paradoxically, it does not define the result to be reached, any more than it creates any obligation of result. It confines itself to strengthening the obligation of means. All these confusions are evidence of the difficulties of the changes to be carried out in the modes of production of knowledge as well as in the modes of instruction and decision making.

Precaution and Responsibility

Those who have made a name for themselves defending and illustrating alarmist conceptions of the precautionary principle are not content with emphasizing what they see as the grave risk of decision-making paralysis. They add that a radical transformation is to be feared in the regime of responsibility of decision makers. According to them, the development of precautionary approaches would lead to an increase in legal challenges to political and administrative decisions. They reckon that the precautionary principle is invoked today more to avoid finding oneself before a judge than for the protection of consumers.[20] If health professionals emphasize the risks of a misuse of precaution understood as a shield, it is because they fear it becoming a resource for those who see themselves as the victims of a lack of prudence.

The view of the French Supreme Court mentioned above emphasizes the advance in Western law of a "theory of victimization" according to which every individual struck by an unfortunate event would see himself as a victim of society deserving compensation. This assimilation, which the Supreme Court considers dangerous, increases the confusion between risk and fault. In fact, a debate has opened up on the degree to which reference to precaution would lead to an extension of the notion of fault, both in order to impose sanctions on behavior for lack of vigilance and to obtain compensation for harms suffered against which solely preventive considerations could not provide sufficient guarantee. The question is not without basis. In the context of a precautionary approach, uncertainty does not mean exemption from responsibility. On the contrary, it strengthens responsibility by creating a duty of prudence. But to date we have no example in which precaution can be said to have modified the system of responsibility. In the area of health, compensation for medical accidents is always granted by reference to responsibility for risk. And in the penal trials that followed the drama of infected blood, the legal proceedings bear on the facts that come under an absence of prevention (delay in establishing the test once the risk was known) and not an absence of precaution during the phase of uncertainty. With regard to the environment, the few decisions made essentially show that precaution has been invoked to challenge the validity of administrative decisions. Moreover, it should be noted that reference to the precautionary principle does not operate in a unilateral manner. In the few disputes that invoke the principle it is as much an excess of precaution as its insufficiency that has been attacked. Despite certain fears, the introduction of the precautionary principle has not disrupted the traditional system of accountability. It has extended it without changing its main components, and the courts (civil and administrative) tend to interpret it with moderation.

The French administrative courts generally leave the administration a wide power of assessment on the appropriateness of measures; following the same direction, the Court of Justice of the European Community has dismissed actions for abuse of precaution on two occasions, judging that there was no manifest error or exceeding of the power of assessment. On the other hand, the French Supreme Court has shown itself to be stricter when it comes to compliance with procedures imposed out of concern for precaution (content of dossiers, committee advice). On this basis, with regard to GMOs and the demand of Greenpeace, it pronounced a stay of execution of the decree of 5 February 1998 authorizing the cultivation of transgenic corn.

To date, the cases of a pursuit of responsibility on the basis of a failure to comprehend or misuse of precaution are very rare. According to the Kourilsky-Viney report, the principle "seems to have almost never been used *expressis verbis* to justify or dismiss a legal responsibility invoked in a court."[21] This is understandable inasmuch as, to say the least, the plaintiff would have to provide proof of a context of uncertainty regarding a danger deserving vigilance and in order to do this produce the existence of knowledge and observations that have been carried out. No doubt he would also have to prove that proportionate, technically possible, and economically viable measures could have been taken. If the accumulation of these conditions allows us to envisage responsibility lawsuits, they would nevertheless be very tricky to conduct and would give rise to considerable judicial controversy. That is to say, at present their chance of success is still highly questionable. On the other hand, specialists think that the diffusion of reference to precaution may function indirectly by giving a broader or more precise meaning to classical notions like "imprudence." Once again, the practices that can be observed today are very far from confirming the alarmist anticipations of some actors.

Precaution does not fix substantial objectives to be reached. It frames procedures for the evaluation and management of overflows which could occur from the implementation of certain projects. Although this is not their explicit purpose, these procedures, as we will see, aim to foster the double exploration characteristic of hybrid forums.

Precaution as Measured Action

12 December 1999: The oil tanker *Erika* is wrecked off the coast of Brittany. 21 December: A note from the Rennes anti-poisons center warns the

Direction département de l'action sanitaire et sociale of the carcinogenic character of the presence of polycyclic aromatic hydrocarbons. Beginning 25 December, 2,000 people, many of them volunteers, do their best to get rid of the traces of the black flood. 27 December: The association Robin des Bois (Robin Hood) reveals that an assessment carried out on the fuel of the *Tanio*, shipwrecked in 1980 with an identical cargo, concluded that there was a health risk. This news provokes a shock. Of course, safety instructions were widespread among the cleaners (wear gloves, a mask, and goggles), but nothing had been said about the existence of this health risk. Two months later, anger directed at the new bandits of the sea and condemnation of the shipowners' cynicism is turned against the public authorities. Public and private assessments confirm the highly toxic character of the product. They add that, in view of the conditions of exposure, the health risk to the rescuers is minimal. Registration of the voluntary helpers and medical follow-up are established nonetheless. The debate is not closed, however. The volunteers claim they have been "had and manipulated." Some already talk of a legal complaint for putting lives at risk. The government pleads guilty. It acknowledges having been informed of the danger, but it argues that, the risk of damage to health having been judged weak, it "did not know how to present the information." For a voluntary veterinarian, the answer is clear: "If there was a doubt about the carcinogenic risk, the precautionary principle should have come into play." We turn now to this question of the point at which precaution comes into play and the modes of action presupposed by such an approach.

Precaution gives rise to a decision-making dynamic that modifies the relations between science and politics, both in the links between them and in their respective authority. It moves away from the classical schema that drastically separates the time of knowledge from the time of decision. It connects them in a to-and-fro movement that is called upon to continue for as long as uncertainty remains. In the classical schema, scientists tell the truth and establish certainties, and then politicians draw "the obvious conclusions," that is to say, in concrete terms, transpose the analyses addressed to them into decisions. In practice, the supposed superiority of political legitimacy due to elections is obviously subservient to the scientific legitimacy of the experts consulted: It is science that enables uncertainty to be removed, and political authorities are dependent on it. The space of choice left to the politicians is generally reduced, and decision is often the result of the strictly technical analysis of an issue. The political decision is therefore only apparently autonomous; its basis is scientific legitimacy. It is precisely this temporal linking and this fitting together of legitimacies that the process of precaution transforms by permitting the double exploration of problems and identities typical of dialogic democracy. Provided it is rid of the ambiguities that some like to maintain, and of the interpretations partially emptying it of its substance, the precautionary principle promotes a conception and practice of political decision making that corresponds to the dynamics of hybrid forums, as described in the preceding chapters. That is why the detour via an examination of the conditions of application and implementation of this principle will enable us to more clearly define the conception of political decision making through dialogic democracy. The precautionary principle is a driver of action, but progressive action, fed through feedback and debate. This action develops in three separate but correlated dimensions. It requires a warning system, a deepening of knowledge, and temporary measures. Each of these dimensions designates particular actors with specific modes of action and incurring a precise type of responsibility. A definitive, clear-cut decision is replaced by a series of "small" moves in all three dimensions—in other words, small decisions, each of which constitutes an advance but none of which leads to irrevocable commitments. The solemn and dramatic scene of the decider making a clear and irrevocable decision is replaced by a long process, gradually producing a common world which is both desired and tested.

Experience gained in the application of the precautionary principle enables us not only to change our usual perception of the decision-making process but also to characterize better the situations in which dialogic democracy is more appropriate than delegative democracy. That is why, before describing these three registers of action, we will show how the actors, and jurists in particular, have provided certain elements for understanding in which cases and on what terms precaution is recommended. Although this reflection on the precautionary principle as a modality of political action furthers our understanding of how hybrid forums can contribute toward the decision-making process, the opposite is also true. What we have learned about hybrid forums, and especially the procedures that they require in order to function satisfactorily, will help us to remove certain ambiguities on the precautionary principle.

The Field of Application of the Precautionary Principle: A Carefully Delimited Framework of Action

From its first formulations, precaution comprises two completely interdependent dimensions: action and framing. Thus it is contrary to the traditional attitudes of denial and panic.

In a situation of uncertainty regarding the reality of dangers and suspected overflows, the precautionary approach affirms the absolute necessity of action. Furthermore, it defines the general framework in which these actions should be undertaken.

However, the concern for clarification is not constant. Some texts mention the need to take precaution into account with regard to consumption, health, or the environment, without taking pains to make the notion clear. This is the case, for example, in the Maastricht Treaty of 1992, which, while defining the principles that must orientate the environmental policies of the member countries (article 130-R-2), made precaution one of the dimensions of a sustainable development[22] without giving further clarification on the meaning to be given to either of these notions. Fortunately there are other references, and we can argue on the basis of these, because they endeavor to orientate the operational implementation of precaution. A text of the European Commission clearly sets out the existence of a "large space of application of a principle of reasoned precaution." This space is situated between a floor defined by the classical conception of prevention (not to prohibit a product or a procedure so long as the existence of a danger has not been demonstrated), and a ceiling defined by an absolutist conception of precaution (prohibition of any procedure or product so long as their harmlessness has not been demonstrated).[23] In the more recent text of the EC itself, precaution is presented as a "reasoned and structured framework of action enabling scientific uncertainty to be remedied."[24] Four elements of framing define the space of precaution: uncertainty, potential damage, effective measures, and tolerable cost. (See box 6.2.)

The first element of framing concerns the existence of a situation of uncertainty. All the definitions refer to it, but none indicate how it is identified and revealed. The terms used are very broad: absence of a relation between cause and effect, absence of indisputable scientific proofs, and so on. This vagueness generates considerable difficulties for defining the start of the precautionary approach. As we saw in the first chapter, classically a situation of uncertainty is thought to exist when dangers of overflow are suspected, without it being possible to define exactly either its characteristics or its conditions of appearance. Obviously, no statistical modeling is conceivable in such cases. The probabilistic approach requires prior knowledge of the emergent event. It cannot be carried out if the latter has causes and modes of development which are still unknown (a new factor of danger such as HIV, BSE, or the H5N1 virus), or if it appears to rest on causal chains and interactions which are still poorly delimited (as in the case of global warming and gas emissions in the 1990s, nosocomial diseases and

Box 6.2

Reference definitions

Declaration of the international conference on the protection of the North Sea (London, November 1987):

In order to protect the North Sea from possibly damaging effects of the most dangerous substances, a precautionary approach is necessary which may require action to control inputs of such substances, even before a causal link has been established by absolutely clear scientific evidence.

Rio Declaration of 1992, principle 15:

In order to protect the environment, the precautionary approach shall be widely applied by States according to their capabilities. Where there are threats of serious or irreversible damage, lack of full scientific certainty shall not be used as a reason for postponing cost-effective measures to prevent environmental degradation.

French Constitutional Charter of the Environment, 28 February 2005, article 5:

As soon as damage [to the environment] is recognized which could affect the environment in a serious and irreversible manner, even though it might be uncertain in the current state of the scientific knowledge, public authorities should monitor, by the application of the principle of precaution in their relevant domains, the implementation of risk assessment procedures and the adoption of proportionate, provisional measures in order to prevent the spread of damage.

European Commission DG XXIV (consumption, health), December 1998:

The precautionary principle is an approach to the management of risks that is adopted in a situation of scientific uncertainty. It is translated by a requirement of action faced with a potentially serious risk without waiting for the results of scientific research.

---

the circulation of infectious germs, in addition to prior use of antibiotics). Initially, these risk factors can be apprehended only through hypotheses, often lacking the possibility of empirical verification. We are then in a phase of theoretical investigation which brings forward scenarios of knowledge, but without being able to consolidate any of them. To take account of these contexts of uncertainty, and in particular of subsequent, particularly complex contexts in which multiple variables interfere with each other, Olivier Godard employs the notion of "controversial world." He characterizes them by the combination of four variables: competing perceptions of the stakes, a variety of concerned interests (which include absent third parties to be represented, like "future generations"), the degree of reversibility of the phenomena, the degree of consolidation of scientific knowledge.[25] The combination of these four variables enables situations of uncertainty to be differentiated, but without it being possible to rank them clearly by establishing thresholds at which precaution comes into play. We will return to the importance of this first moment of the identification of a danger which enables ignorance to be framed, as it were, and preparation to be made for "initiation of precaution." The other elements of framing are a bit more precise.

The second element concerns a preliminary assessment of the gravity of the suspected danger. Two conceptions of gravity can be distinguished. The first, extensive, leaves a very wide field of assessment; the second, however, is clearly restrictive. In the first group is the 1987 Declaration of the international conference on the protection of the North Sea, which envisages control of emissions of "the most dangerous substances likely to be harmful" to the marine ecosystem. It involves reduction at source of emissions of toxic products which are enduring and susceptible to bioaccumulation. For its part, the European Commission refers to the notion of "potentially serious risk" (definition 4). The second type of definition is clearly more restrictive and pushes the requirements of evaluation of gravity further. This is the case, for example, for the Rio convention and of French law (definitions 2 and 3), which adopt the expression "threat of serious and irreversible damage." The terminology chosen and its redundancy clearly indicate the desire to limit precautionary activities to the most threatening situations. The initial evaluation of the danger and the first expert assessments on the construction of hypotheses of the risks likely to be generated are therefore crucial here. The meaning of these levels of gravity is, of course, constructed in practice. France's precautionary measures against BSE are attributable to the formation by the Dormont Committee (set up by the French government in April 1996) of the collective conviction that BSE could be transmitted to humans. The question of crossing the species barrier was the main assessment criterion of the gravity of the danger; it was in order to provide some elements of an answer that research on the routes taken by non-conventional transmissible agents (NCTA) in general and by prions in particular was launched.

The third element of framing concerns whether initiation of precaution should be optional or mandatory. Definitions 1 and 2 fall under the first, optional model. The convention on the North Sea says that the precautionary approach "*may* require action," which presupposes an assessment and/ or a political debate on the appropriateness of the actions to be taken. The Rio convention introduces another criterion by making action conditional on the capabilities of each state, but no complementary clarification is given with regard to what type of resources reference is to be made. Here too the criteria of assessment are very broad and there is no indication of who is in a position to pronounce on the requisite capabilities. Conversely, other texts judge that the observation of a danger and a first evaluation of possible damage must make protective action mandatory. This is the case for definitions 3 and 4—the French text speaks of "not delaying the adoption of measures," and the European Commission speaks of the "requirement of action faced with a risk." Precaution here is no longer optional; it becomes mandatory.

The fourth element of framing concerns the extent of the measures to be adopted. The proposals vary in terms of the intensity of the action that is expected. The European Commission text gives no indication on this question, thus leaving the field widely open to assessment by public authorities. The convention on the North Sea is content with calling for limitation of the risk factor. The Rio convention introduces a double criterion by adopting "cost-effective" preventive measures. The reference to prevention allows us to envisage severe measures aiming to prevent overflows. But the term 'prevention' is inadequate here, for it presupposes a danger objectified as risk, whereas 'precaution' refers to a context of uncertainty and 'controversial worlds' in which measures have not yet been taken. The second criterion used is clearer, but much more restrictive since it makes the measure to be taken depend on a cost-benefit analysis. Finally, the French formulation is the most precise, inasmuch as it adds two further criteria to those of the Rio convention, which it takes up. We find again the notion of the cost of preventive measures having to be economically sustainable. In the first place it supplements these criteria through the requirement of effectiveness, that is to say, through a responsibility of the decision maker with regard to the implementation of the measures taken. The memory of the dramatic problems raised by the failure to respect measures for the selection of donors in the affair of infected blood was no doubt in the legislator's mind. But the requirement of impact is tempered by the last criterion, that of proportionality of the measure to the envisaged risk. It involves a limitation that supplements that of acceptable cost. For there may be measures of low cost economically (like halting a vaccination or some preventive examinations), and so relatively attractive to decision makers, but which, through their radical character, would be out of proportion with the risk they aim to eliminate. Thus the policy of slowing down vaccination against hepatitis B was often criticized by professionals as a measure with no relation to its supposed neurological effects (multiple sclerosis). Here again, the decision to initiate the precautionary approach and the choice of its first modalities presuppose precise clarifications on the initial evaluation of the danger.

All these framing efforts carried out by national and multinational agencies pursue two main objectives. On the one hand, they want to orientate collective action by specifying the field of application and the concrete modes of implementation of this new norm of decision making. In fact, the public authorities are regularly led to adjust divergent interests like those that pit the defense of individual freedoms against the guarantee of collective security, which is translated here as the search for a combination between free enterprise and the need to reduce the impact of negative effects on the environment and human health. This type of combination is particularly delicate in situations of uncertainty and requires particular guidance. On the other hand, these guidelines also want to prevent certain abuses and in particular to suppress the discretionary use of this notion in international exchanges for protectionist purposes. This was the reproach the British beef farmers directed at the French authorities when the latter opposed lifting the embargo in September 1999.

Let us summarize. The precautionary principle is applied in situations in which uncertain but grave dangers are plausible; it requires effective and economically sustainable measures to be taken to avoid the materialization of these dangers. As we will see later, each of these measures presupposes a public debate. The initiation of precaution and the measures to which it leads need the space of hybrid forums.

All those who have taken part in the still-incomplete elaboration of the precautionary principle have not confined themselves to reflecting on its field and framework of application. They have been equally concerned with the modes of action that give it a concrete existence. Precaution is not synonymous with non-decision and temporization. It is embodied in approaches, indeed in apparatuses, of which we will make an inventory closely related to the experiments realized by the actors. We will thus see the emergence of a new conception of decision making at the heart of dialogic democracy.

The Initiation of Precaution: Vigilance, Exploration, and Choice of Measures

A precondition of precaution is ascertainment of a situation of uncertainty that is likely to cause grave damage. The point of departure of the approach is the identification of potentially negative effects arising from a phenomenon, an activity, or a product. Depending on the case, such a context either may or should trigger effective actions that are both proportionate and economically sustainable. The importance of this guideline as regards the environment as well as the accumulation of recent tests with regard to health safety enable us to clarify the major question of the practical modes of implementing precaution. The preliminary stages of the debate on the "nature" of precaution or on the identification of its addressees are now out of date. It is by moving forward in the modeling of induced ways of action that precaution will become clearer as a new benchmark for decision making.

We will show first that, contrary to analyses which fear that recourse to precaution will direct the management of dangers toward irrational practices, this model increasingly tends to be operationalized in ways inspired by the classical model of risk evaluation. Yet in our view this development should be challenged. Adopting the perspective opened up by the previous chapters, we develop instead a gradual approach in which actors and lay knowledge should be integrated as soon as possible in the activities of vigilance, exploration, and the choice of measures to be taken.

From Vigilance to Alarm   Precaution is possible only when an empirical or institutional system of vigilance exists, that is to say, a more or less formalized set of socio-technical arrangements that enables the collection, recording, and collation of information which, while dispersed and heterogeneous, is likely to reveal a broader collective problem. This is the point at which dangers are identified, which is the phase prior to the alarm strictly speaking.

The report of real but still-unexplained damage authorizes us to suspect the existence of a biological, chemical, or physical agent, likely to have an adverse effect on human, animal, or vegetable health, or on the balance of the environment. The starting point of the perception of danger is observation of a symptomatology; the identification of a complete etiology comes into play only later. The situation of uncertainty stretches between these two moments. Thus, it was possible to express a fear of an epidemic of listeriosis in France at the end of November 1999 on the basis of observations which were not very extensive quantitatively, but which were enough to mobilize the health authorities.[26] The main sources of information were the regional health observatories and the network of family doctors. Today, national sanitary surveillance institutes exchange data on avian flue in humans and animals on a daily basis with the World Health Organization. In contrast, the controversy that broke out again in 1997 on the risks of leukemia linked to closeness to nuclear waste disposal at La Hague, in

Normandy, revealed the absence of any system of epidemiological follow-up around the power plant. The researchers, the authors of the report, emphasized the difficulty they had in locating possible pathogenic effects inasmuch as there had been no regular follow-up on the state of health of the population since the creation of the establishment. Therefore, as far as it was possible, they had to reconstruct the indicators that were necessary for their approach. In other countries, like Great Britain, arrangements for follow-up checks on the state of health of the population of regions affected by nuclear plants accompanied their installation from the start, indicating a completely different policy of attention to risks.

To describe the complicated process that leads some actors to detect what they see as warning signs in the flux of events, Chateauraynaud and Torny propose to focus on activities prior to the alarm, to what they call "attention-vigilance."[27] These actions in the face of uncertainty rest on a perception of dangers, on the social actors' capacity for attention, which sometimes arise from previous real-life trials in confrontation with a risk. They are upstream of formalized alarms and disputes, and their dynamic oscillates between "disquiet" and the "collection of information to keep tabs on phenomena linked to the most everyday activities." This "attention-vigilance" is most often linked to the immediacy of exposure to danger and the absence of a satisfactory interpretation which would enable one to understand it and protect oneself from it. We referred above to the case of the popular epidemiology practiced by the inhabitants of Woburn and the shepherds of Sellafield. We can take another example from the Minamata catastrophe in Japan. Beginning in 1953, in the fishing villages at the mouth of a river, pregnant women gave birth to children with monstrous deformities. In the end there were close to 1,500 victims, more than a third of whom died young. Another 5,000 people were affected to a lesser degree. The inhabitants accused a metallurgical enterprise, Chisso, which was situated some kilometers upstream and had always discharged its waste into the water. The pre-existence of substantial chronic pollution gave the inhabitants an interpretative framework for the damage and for imputation of responsibility. Initially, the enterprise denied any link between its activities and the health catastrophe, and it continued with its discharges. It took four years to understand the origin of the observed natal malformations. In 1957, after exploring a number of hypotheses, a commission of official experts established their origin in mercury. Methyl mercury was found in strong concentrations in the blood, livers, and brains of the inhabitants. It had irreversible effects on the embryos. But how was this mercury absorbed by the mothers? It was another two years before the complete cycle of contamination was finally demonstrated in 1959. Among other substances, the enterprise dumped mercury into the river. The mercury reached the sea and accumulated in the depths where it impregnated the plankton on which the fish caught at Minamata, the regular food base of the population, fed. The intuitive analysis of the Minamata fishermen turned out to be right, although their model of interpretation of the situation was incomplete. They incriminated the water, which was only a vector of transmission. However, their closeness to the source of risk, as well as with the symptoms produced, put them in a position of attention-vigilance that could have helped reduce the effects if it had been taken into account. Social receptiveness to the networks of proximity of initial perception of dangers is certainly one of the contributions of the precautionary approach. It leads to a consideration of information linked to a more concrete than theoretical perception of threats.

Attention-vigilance, which passes through the consideration of new information with a limited audience, leads to the renewal of pre-existing frameworks of reasoning by enriching them. In this sense it corresponds well to the initial phase of the precautionary approach. Thus it has been possible to show that in the progressive discovery of the BSE epidemic, the relative quickness with which specialists of STSE-type[28] diseases were mobilized is explained by, among other things, their continual vigilance since the 1960s with regard to a family of diseases caused by non-conventional transmissible agents. Furthermore, the spread of information to the broad public on the epizootic, and more particularly the transmission of images of the sick animals, stimulated the attention of the farmers, whose role was as important as that of the veterinarians in making the disease visible.

The identification of potentially negative effects is thus produced by hybrid networks in which the professionals theoretically in charge of the problem are not necessarily in a central position. Laypersons and their "epidemiology in the wild" often occupy a decisive place through their ability to make connections between empirical observations and general information. The mobilization and activism of some of them succeed, sometimes, in breaking the complicity of the economic and professional interests that strongly deny some dangers. Thus, after an initial crisis at the end of the 1970s, it took the painful journey of workers who were victims of asbestos to shake the common front of industrialists, company doctors, and specialists of pulmonary diseases and so arrive at complete prohibition of the product in February 1996 and the self-criticism on the part of some of these specialists. In fact, good exploration of a danger requires the active participation of the threatened populations, always within the limit of some of them withdrawing from the collective out of fear of "being transformed into guinea-pigs." Entering into the logic of action of the precautionary approach therefore means creating the conditions for collaboration between specialists and laypersons in the networks of vigilance.

There is no initiation of precaution without the identification of a danger, without locating damage. These actions therefore constitute the first stage. But to go further in the implementation of precaution, they must be accompanied by a "first evaluation," by an exploration of the overflows and their extent. This approach has nothing to do with the traditional apprehension of already delimited risks. The determination of the threshold of activation permitting transition from vigilance to alarm, from discovery to the first temporary measures, passes by way of a work of investigation and metrology that we will not consider now.

The Exploration and the Measure of Overflows   Starting from the initial identification of a danger, precautionary practice requires a preliminary evaluation of the overflows and associated dangers in order to assess its gravity. This exploration must include analysis of the nature and extent of the danger, its possible causes, its modes of diffusion, and its factors of sensitivity. It involves assessing how much it is possible to fear that a potentially dangerous effect for the environment, for human, animal, or vegetable health, is incompatible with the level of protection deemed desirable.

In justifying the embargo against British beef, the European Court of Justice based its decisions precisely on a prior assessment in order to validate the measures taken. Clearly there is nothing irrational about these measures; they were seen to be legitimate because they were preceded by an exploratory approach. In its decisions of May 1998, the Court thought that the information available indicated that the risks should be seen as potentially serious. It also reckoned that the existence of about ten atypical diseases made the "theoretical hypothesis" of transmissibility to humans credible.

Lack of certainty does not in fact mean a complete absence of knowledge. Studies have reconstructed the history of the mad cow disease in order to try to see how, and with what effects, the precautionary principle could and should have been implemented. They recall the existence of a whole set of things that were already known, both about NCTA-type infectious agents and about intra- and inter-species transmissibility. Moreover, it was on the basis of this knowledge that the hypothesis of transmissibility to humans was finally accepted.[29] They also emphasize the paradox of the origin of dissemination of the epizootic, which is attributed to a change in the technique of production of animal feed at the end of the 1960s. The new process was supposed to be economically more efficient, but also more ecological (abandoning the use of solvents, less energy consumption). But, they note that a double-sided evaluation of risks and advantages was not undertaken: "If the evaluation had been made we would have come across the knowledge just described," and in particular the fact that NCTA were not inactivated by the new methods, that the oral route could be an effective route of infection, and that the feeds therefore had an "amplifying and disseminating" nature.

In terms of the exploration of danger, the authors draw several conclusions of general significance from this example. In the first place, in their view attention should focus as much on the processes as on the products, and the search for information should be pursued within the most diversified practical horizons. This presupposes prior organization of the traceability of actions so that the detailed sequences of operations linked to the situation identified as dangerous can be reconstructed. Then, exploration should weave together the dispersed and heterogeneous information in order to construct "bundles of convergent indices." The objective is not to find *one* consolidated and replicable proof, but the gradual construction of hypotheses, combining theoretical data with empirical observations, objective and subjective data. The World Trade Organization agreement on the circulation of health and phytosanitary products thus allows for exceptions to the principle of free exchange. In situations of insufficient scientific proof, countries can take provisional restrictive measures while waiting for "a more objective evaluation of the risk." This formulation allows the inference that, *a contrario*, precautionary measures may be supported by a more subjective evaluation of danger. In the same way, in the meaning given to the initial evaluation by the European Commission, exploration may use "nonquantifiable data of a factual or qualitative nature" and not be limited solely to statistical data. This attention to qualitative sources also finds expression in the attention accorded to views formulated by minority fractions of the scientific community. As a rule these views are ignored, while they often warn against dangers inherent in the translations carried out by secluded research. The precautionary approach leads to these views being taken into account, for they are seen as revealing uncertainties which are underestimated by most researchers.

How can all these positions be brought together without leaving aside those that are the most heterodox? There is a strong temptation to be satisfied with an enriched expertise, but an expertise that still does not really advance dialogic democracy. The experiences of different committees that have had to deal with situations of uncertainty show the force of the immediate desire for "more science" in order to realize "the best possible expertise." Thus most specialists reckon today that the establishment of a counter-expertise to supplement and enrich classical expertise is sufficient to ensure the diversity of exploratory paths. The creation of governmental agencies specialized in health evaluation illustrates this approach. They are often put forward as offering a sufficient guarantee of pluralism to ensure the diversification of analytical and evaluative frameworks. These agencies are often content to take up the classical models of decision making and consequently are confronted with the difficulties generated by these models. This is why the idea of recourse to a pluralist expertise in situations of uncertainty, bringing together not only specialists of different disciplines, administrators, and decision makers, but also, and above all, different categories of laypersons, is gaining ground today. The orientation document of the Directorate of Consumption and Health of the European Community recommends, for example, the introduction of "transparent" procedures in the event of potentially serious danger, involving "all the parties concerned at the earliest possible stage."[30] The document is still very timid, however, for it only envisages being open to lay opinions under a doubly restrictive point of view. In the first place, it limits their intervention to the study of diverse options in the management of the danger, once the initial exploration has been undertaken. Then, the main reason invoked to justify their participation is its contribution to the legitimacy of measures which are not entirely based on science. Recourse to a pluralism of points of view is reduced to a strictly utilitarian function; it is supposed to ease the way to the famous social acceptability, the limits of which have been shown by the actors of hybrid forums.

What emerges from previous chapters is that to avoid creating a discrepancy between the measures implementing the precautionary principle and the dynamic of dialogic democracy, there must be a very early opening up to and confrontation between points of view at the point when the first information is gathered. The analysis of dialogic procedures suggests that minority or dissident hypotheses outside of the existing frameworks should be expressed and considered when the investigations and research are decided on, and not afterwards. This type of dissenting and unexpected questioning is a perfect illustration of the irreplaceable role of research in the wild that we presented in chapters 2 and 3. The questions raised by GMO, for example, are not all contained in the confined space of the conceptions developed by biologists; they also—first and foremost?—concern farmers, consumers, and defenders of the environment. A similar argument can be applied to the problem of climate change, which is typical of the shifting of questions that may be generated by the gradual introduction of marginal questions into the most official research. Since the 1970s, what to start with was only a millenarian prophecy close to the catastrophism of deep ecology has become a field of multidisciplinary research that is also an important stake in international negotiations. Today, considerable means of observation of the globe and the atmosphere are enlisted by the most prestigious scientific bodies to evaluate the probability of risk and the means to prevent it.[31]

The greatest attention should be paid therefore to the design of the arrangements for gathering and handling information and points of view. There is a strong temptation to reduce them to no more than the extension of existing structures. From this point of view, reference to the notion of expertise is dangerous. Allow us to recall the wisdom of the proposals made by all those who are working on the organization of hybrid forums. The lesson must not be lost. We have seen that to facilitate the discovery of a common world it is essential to create the conditions for the to-and-fro movement between the two explorations of possible worlds and of the collective. The precautionary approach must not impede the dynamic of explorations.

The greatest firmness is needed to avoid this danger. The aim is not to arrive at definitive, clear-cut decisions at any cost. The model of action is that of measured action. The polysemy of the expression "taking measures" invites us to recognize that the challenge is to make it possible to measure (in the metrological sense) overflows so that measures (in the political sense) can be taken to contain and control them. It is a matter therefore of fostering the differentiation of two moments in the exploration and evaluation of dangers, instead of forcing them into unified structures of expertise. The first stage aims to take the measure of the damage and to redistribute the zones of uncertainty between those already located and those that are gradually being discovered. The second stage is the assessment of threats, and it leads to precautionary measures on more assured bases. During the first stage, the consideration of empirical data, some of which are marked by the subjectivity of lived experience while others are based on atypical theorizing, should not be seen as a simple palliative for the insufficiency of data which can be modeled. On the contrary, their integration in the collective reasoning to which the exploration gives rise should have weight equal to the "objective" data in working out scenarios and hypotheses. In any case, the time of exploration should give rise to an intense circulation of information and favor the multiplication of interpretations and hypotheses, whether their origin is scientific, professional, or lay. As the previous chapter showed, apparatuses working in parallel, bringing together homogeneous actors and questioning each other, are more productive than big agencies mixing heterogeneous points of view and in which traditional problematizations always tend to impose themselves.

As we have said, the initiation of precaution requires an evaluation of the seriousness of the dangers incurred. This evaluation is backed up by the measure of the overflows, and to be as robust as possible this measure can be realized only within the framework of structures that enable the actors, whether experts or laypersons, to openly associate with each other. Precaution is jeopardized if this openness is forgotten to the advantage of structures of expertise which, even if they are diversified, are ill suited to the double exploration of possible worlds and of the collective.

The Choice of Measures Measurement of overflows in order to take measures enabling them to be controlled is the necessary condition for the initiation of precaution. But how are the measures to be taken worked out and how do we choose which ones to take? Since precaution takes us away from the traditional models of decision making, we have to redefine the appropriate criteria.

The precautionary principle does not lay down "a ready-made model of management." Definition of the measures to be taken, their adaptation to the situations to which they are applied, as well as their follow-up, give rise to constant polemics. This was the case, for example, for the decision by the French government in October 1998 to limit vaccines against hepatitis B on the grounds of a suspected risk of triggering multiple sclerosis in young children. This measure provoked the lively reaction of public health specialists, who considered it very inadequately based. It had a massive effect nevertheless. Between 1996 and 1999, the number of vaccinations was divided by nine. In March 2000, a report from experts excluded the existence of a high risk without dismissing a low risk for populations with particular factors of sensitivity. Owing to an inability to improve medical checkups in schools, the shortcomings of which largely explain the initial decision, the restrictive measures were not revoked. Thus, the choice of measures raises several types of questions: When should the choice come into play? What should the scale of the measures taken be? How can their implementation be guaranteed? How can we ensure that lessons are drawn so as to enlighten the decision makers? The first two points aim to avoid the adoption of arbitrary or disproportionate measures. The third and the fourth are of a different order. They lead to questions about the conditions of application and of the follow-up of measures based on unusual justifications.

Determining the best moment to implement precautionary measures is obviously very tricky. If taken too soon there is the danger that they will not contribute to the dynamic of exploration that enables knowledge to be amassed and points of view to be gathered; taken too late, they may prove to be powerless in the face of worrying overflows. The experience of the different crises that the political authorities have had to manage in recent years suggests that implementation should be activated when, confronted with a potentially dangerous event, exploration reveals limits in the existing knowledge: a symptomatology without etiology (partial identification), a link between dose and effect not established or considerable variability of the pathogenic agent (poorly delimited adverse effects), uncertainty as to the factors of diffusion or reception (poor assessment of exposure). In these conditions, the overflow seems to be defined, or at least strongly suspected, while there is maximum uncertainty about its description and origins. Such a situation entails taking the danger seriously and adoption of the most pessimistic hypothesis. This is not the "hypothesis of the worst case"; rather, it is situated midway between underestimation and overestimation.

As for the criteria of the choice of measures, these correspond to three requirements. The measures taken must first take account of the probable development of scientific knowledge still in its infancy, in order to reduce the risk and avoid later stricter decisions. Second, they must take into consideration the possible medium-term (a long period of incubation of a pathology) and long-term effects (genetic mutations, endocrinal disturbances linked to bioaccumulations of toxic or radioactive substances, irreversible transformations of ecosystems). In this sense, there is agreement today that precautionary measures must not only be inscribed within a perspective of sustainable development but also must be attentive to questions of equity between generations. This has implications for their scale. Contrary to what a superficial reflection might lead us to think, precaution cannot be identified with gradual measures which have a limited impact at the start and can then be strengthened after the precise evaluation of the risk. The opposite approach must be followed. Drawing on the examples of nuclear power and aeronautics, Marie-Angèle Hermitte was the first to see that overestimation of the scale of measures was a major element of a precautionary policy.[32] She shows that to be fully effective and to limit exposure, the precautionary measures taken right at the start in the case of BSE would have required a much larger scale: "The greater the uncertainty, the more it is necessary to act on a large scale, even if this means gradually reducing the precautions with the improvement of knowledge."[33] Slaughtering all animals that have had the same exposure to the risk (that is to say, having had the same feed), and not just the sick animals, is an example of this. The scale of measures raises directly the question of their economic cost. This can only be resolved case by case within the framework of consultations that allow for dialogic procedures. According to a strict economic logic, the annual investment of 110 million francs to screen for HIV in every blood donation is, according to some experts, a disproportionate measure (it would enable half a life to be saved every 20 years), whereas the effect of allocating the same sum for cancer screening would be 200 times better. Even so, in the context of the crises linked to HIV, this measure was undoubtedly proportionate to the result sought, that is to say, restoration of confidence in blood transfusions and the search for the lowest possible risk of infection. Another recent example is provided by the precautionary measure taken by Denmark at the beginning of March 2000. A directive from the minister responsible for food products asked distribution outlets to suspend the sale of beef after the discovery of a case of mad cow disease in the north of the country was announced. Producers and liberals launched a huge polemic, reckoning the measure to be disproportionate and ruinous. The government stuck by its decision, considering the protection of the health of consumers to be an absolute objective and that it was also the kind of measure needed to restore confidence and preserve both internal and export markets. In the end, only a political authority can assume responsibility for balancing interests in such cases.

The question of the design of measures also involves the analysis of the conditions of their implementation. This is the third criterion. Today this is considered to be the determining point. The choice of measures for implementing a public program of action (selection of blood donors, prohibition of feed coming from cattle) is too often made without considering the concrete functioning of the organizations involved in the implementation. Their constraints, their specific objectives, and their representations of the problem and ability to deal with it are not really taken into account in the choice of the measures carried out. Now, these factors immediately come into play when it is a matter of carrying out the interpretation and adaptation of the measures. More generally, the question of the measures that can really be taken in a context of uncertainty must be posed. As proposed by Hermitte, the actions that accompany the measures should be the object of consultation and exchange of information involving different addressees and, in particular, the actors responsible for implementation. There is no doubt that recourse to dialogic procedures could make a difference. At the least, divisions and resentments would be reduced.

A New Conception of Decision

Apart from remaining ambiguities and contradictions, the work and reflection devoted to the precautionary principle, considered from the point of view of its contribution to dialogic democracy, have helped to produce a new conception of political decision. This is not happening without gnashing of teeth and violent polemics.

Among the many debates that implementation of the precautionary approach gives rise to today, three merit brief discussion. The first turns on the difficulty of escaping the notion's polysemy. Precaution clearly has different meanings and takes different forms depending on the contexts in which it is invoked. Thus, with regard to medical treatment (the use of chemical substances, surgical intervention), precaution is inseparable from the costs-benefits assessment carried out in the interest of the person being treated. This may lead to the acceptance of interventions allowing one to reckon on a vital gain or improvement of the quality of life, even when they are likely to produce damage that is difficult to specify. On the other hand, precaution should be more constraining when the expected gains concern only one category of actors, while the damage may concern the majority. Thus, with regard to the cultivation of GMO, if the economic gains (productivity) are known, the other gains (possible improvement of the quality of the product) are yet very uncertain, as are also the negative consequence on the environment and public health. These differences, which stem from constellations of opposing interests, exist. But dialogic procedures are designed precisely to take them into consideration in the choice of measures to be decided.

The second major difficulty which some have identified is that of the problematic return to confidence after the first measures taken are subsequently shown to have lacked an object. Some economic actors, for example, fear that a precautionary approach may definitively damage the image of a product or activity and that it may be impossible to restore the credit lost. However, the obligation to display the colorings and preservatives used in consumer goods, or the revival of the market for beef after the fall in sales linked to the announcement of BSE, suggests that we temper this judgment. We also note that in 1998 a product equivalent to Distilben was authorized for sale in the United States after the composition of the product had been modified to avoid the hormonal risks of its initial version. The producers were smart enough to rename this drug.

The third difficulty is the latent contradiction between the requirement of the effectiveness of precaution and that of the democratization of scientific and technical choices, which is the main theme of this book. This is a major difficulty. Unlike the previous two, it is a contradiction that is not easy to overcome. Organizing the explorations, debates, and consultations that hybrid forums presuppose requires time. The requirement of precaution, on the other hand, calls for the earliest and most radical intervention possible. The warning must create a situation of urgency, and Hermitte develops the idea of a law of crisis "imposing special obligations within the framework of precise procedures."[34] In such a context, setting up hybrid forums is far from obvious. But whatever the urgency, it remains the case that all the activities that materialize precaution—evaluating and broadening the available information, defining the populations exposed, weighing the costs and benefits of dangerous activities, choosing the measures targeted, and so on—require wide discussion. This means that the procedures adopted to organize hybrid forums must be well identified and running smoothly so that their implementation is speedy and effective.

To conclude provisionally on precautionary approaches, we should recall that the three activities that we have distinguished—attention-vigilance, exploration, choice of measures—do not take place in chronological order. On the contrary, during the time of precaution they interact dynamically. The diffusion of information produced by the first explorations as well as the effects of the first measures are likely to stimulate attention-vigilance or to reorient it toward other networks of actors. The problems of the implementation of the first measures may lead to other exploratory paths, and so on. Precaution, and this is compatible with the general approach of

Table 6.1
Two decision models.

"Clear-cut" (traditional decision)	"Series of rendezvous" (decision in uncertainty)
A single moment, an individual act	A repeated activity linking together second-order decisions
Carried out by a legitimate actor	Involving a network of actors with diverse responsibilities
Closed off by scientific or political authority	Reversible, open to new information or to new formulations of what is at stake

this book, is a process of producing knowledge; it is also an exploration of identities which will eventually make up the collective. It does not define the boundaries of the acceptable and the unacceptable. It involves a type of judgment that, within a rule of action, readily leaves indeterminacy.[35] Such a judgment does not have a univocal meaning which is imposed *a priori* on social actors and determines their perceptions and behavior. It is not a matter of a pre-defined model of action that serves as a norm or general measure for judging actions. With all the more reason, precaution is not in any way a legally sanctioned imperative. No more does it enable one to decide whether an act is in itself just or unjust. As we will see in the next chapter, the soundness of the measures taken depends entirely on that of the procedures followed to take them.

The emergence of a notion like precaution testifies to profound changes in decision theory. The traditional decision rests on the model of the "clear-cut choice," that the individual decider endorses after consultation. In a context of uncertainty, the sequential model loses its pertinence and apparent coherence to the advantage of an iterative model that may be described as a series of rendezvous. Three essential breaks should be noted. First, there is a transition from the singular of the individual act to the plural of repeated activity. Then, an individual decision is expanded to a decision that involves a network of diversified actors. And finally, the clear-cut decision claims to close the case, whereas decision in a context of uncertainty can be revised, remaining open to new information or new formulations of what is at stake. This perception is consistent with what political science has always said: Clear-cut decisions are the outcome of a series of micro-decisions, as are their application. In general, political procedures make only clear-cut decisions debatable, not the flows of micro-decisions which prepare or implement them. The precautionary principle, in contrast, is intended to make all decisions debatable, along with the intermediate results achieved through them. A decision made by a composed collective has neither the same shape nor the same content as a decision made by an aggregated collective. This is summarized in table 6.1.

7 The Democratization of Democracy

Have we done what we said we would? Focusing our analysis and reflection on hybrid forums, we have put forward the hypothesis that they make a powerful contribution to the enrichment of democratic institutions. In fact, when uncertainties about possible states of the world and the constitution of the collective are dominant, the procedures of delegative democracy are shown to be unable to take the measure of the overflows provoked by science and technology. Other procedures of consultation and mobilization must be devised; other modes of decision making must be invented. As we hope to have shown, the innumerable actors involved in socio-technical controversies contribute to these procedural innovations. And if we have been able to reveal these innovations, it is because we are freed from a set of categories and grand narratives that conceal, to the point of making invisible, this anonymous, collective, stubborn work that, day after day, brings dialogic democracy into existence.

It is time to draw up a balance sheet of losses and gains, of relinquishments and profits. It is time to show, first, that we were not wrong, initially at least, in not talking about risk and even more of a risk society, in not holding forth on the notion of expertise, in not taking up the classical dichotomies between culture and nature, facts and values, and in not considering the possibility of leaving uncertainties to the care of economic mechanisms. All these notions and questions have not been spirited away; they take on a more precise and useful meaning when it is shown that they are inscribed in the logic of delegative rather than dialogic democracy.

Once we have freed the terrain and put delegative democracy in its place, it becomes conceivable to pose the general question of the contribution of hybrid forums to the enrichment of the procedures of representative democracy. What innovations are likely to be transposed? What lessons can we draw that can be called upon and deployed outside of hybrid forums alone?

There is obviously no question of giving exhaustive answers to these questions. We will confine ourselves to something that seems to us to be a central contribution of technical democracy: the demonstration that it is possible to find an equitable solution to the insistent question of the representation of minorities.

The End of the Grand Narratives

A Risk Society?

With his book *Risk Society*, the German sociologist Ulrich Beck gained a reputation among those who are interested in the relationships between science and politics.[1] Transferring a notion taken from the vocabulary of engineers, economists, and insurers into the world of philosophy and, indirectly, of the social sciences, Beck identifies a paradox. Science and technology, he observes, constantly produce unexpected and often negative effects. By dint of repetition, these overflows, which specialists are unaware of or refuse to anticipate, end up undermining the scientific institution from outside. How can laypersons and ordinary citizens continue to have confidence in science and its priests if the misfortunes that come from Pandora's Box continue to fall on their heads? Science and technology, and with them scientific progress and, indeed, progress *tout court*, have become the objects of a generalized mistrust. The paradox is that to get out of this situation of suspicion the ordinary citizen has no other strategy than to appeal to the scientists. It is the latter, Beck comments, who have instruments and skills that enable them not only to establish the existence and effects of these overflows but also to find remedies for the misfortunes they cause.

The only rational strategy that remains open to ordinary citizens is that of suspicion. To change the relation of force unfavorable to them, and to force professionals to take account of their fears and explore the overflows brought about by science and technology, laypersons must establish public debates so that the anxieties, fears, and doubts that poison their private lives are expressed. For Beck, the organization of hybrid forums (supposing that the question interests him) would have only one objective: constraining scientists and technologists to come out from their seclusion, both to provide explanations and to take into account facts that they are unaware of or (worse) that they try desperately not to know. The legitimacy of such a trial of strength, through which the citizens' and laypersons' mistrust is publicly expressed, is rooted in the specialists' political and financial dependence. The ordinary citizen finances the researchers by buying the products he consumes and by paying his taxes; consequently, he has a right to control their activities.

Risk society is a society of generalized mistrust founded on a paradox: When a citizen wishes to resolve the problems that the specialists were unable to foresee or to avoid, he finds himself back in their hands! He therefore has no other solution than to maintain the delegation while increasing the mechanisms of control and supervision.

Is a society in which everyone mistrusts everyone else inevitable? The paradox that torments Beck disappears when we reconsider the hypotheses on which it is founded. Specialists, no more than institutional spokespersons, do not have any monopoly. Beck, like many other philosophers and sociologists, tends to consider as taken for granted and non-negotiable the two delegations that hybrid forums precisely endeavor to bring back into discussion. For him, it is the destiny of ordinary citizens and laypersons to be excluded, by definition, and so irremediably, from science and from political representation. This exclusion is constitutive of democracy. It makes the involvement of laypersons in cooperative research, as well as the active participation of emergent groups in the negotiation of their identities, unthinkable. The two divisions are recognized, deplored, denounced, and ... accepted!

By refusing, implicitly at least, to envisage going beyond the double delegation, Beck ends up giving the notion of risk a central role. For the sciences and technologies to be politically controllable in a delegative democracy, we have to accept in fact that, on the condition that they are well controlled and suitably encouraged, researchers and engineers are able to take an inventory of the possible states of the world—in short, to describe the set of likely scenarios. If this were not the case, if the existence of radical uncertainties were accepted, rational debate would no longer be possible and no reasonable decision would be conceivable. Actors would no longer have any choice except between immobility (refusing an uncontrollable progress) and absurdity (leaping into the unknown). By dispensing with the formidable tool of dialogic procedures, which were invented to confront uncertainties and to avoid having to choose between the plague and cholera, Beck is forced to reduce politics to the (social) negotiation of risks and their distribution. In this perspective, what is at stake for actors is not the pursuit of a still-unknown common world but the choice of a world from those which are known or can be anticipated. And this choice (such, at least, is the postulate on which the notion of risk society is based) results from the compromise that is finally reached between actors

who have different assessments of the risks (and especially of their probabilities) and who have a greater or lesser propensity to take a particular
risk. Is it possible to go further in a vision that populates society with individual agents concerned solely with the calculation of their interest?

There is nothing surprising in the fact that engineers, economists, or
insurers privilege the notion of risk; they are pursuing their trade. In fact,
all these professions nourish a profound aversion for uncertainties and
their collective management. Were they to agree to enter hybrid forums,
which raise the question of the modes of exploring possible worlds and
communities, they would be forced to recognize both the importance of research in the wild and the existence of emergent identities which are more
concerned about being recognized than about simple calculations of risk.
And this is what they do not want!

But it may seem strange that philosophers and sociologists take part in
keeping the various attempts to develop and experiment with dialogic procedures in the dark. Hybrid forums, the existence and inventiveness of
which we have acknowledged throughout this book, are simply ignored.
The route we have followed, which has led us to take these experiments as
our starting point, has protected us against this morality of the insurer and
the engineer. Risk is that which remains to be discussed once the work of
exploration of technical and political uncertainties has been taken to its
end. To make it the first and only point of the agenda is to refuse, with a
sort of aristocratic disdain, to take seriously the many attempts by actors
to invent forms of organization of hybrid forums that will enable them to
devise scenarios rather than just choose between scenarios.

Democratizing Expertise?

Another notion, equally omnipresent in the literature, has disappeared:
that of the expert. Much has been written on the nature of the expert and
on the thousand and one ways of organizing and calling upon expertise.
The subject is obviously not without interest. But it is only one minor
aspect of the more general question of the organization of hybrid forums.
It is a point on the map of technical democracy; it is not the whole map.

What is an expert? Answer: someone who masters skills with recognized
(indeed certified) competence which he calls upon (either on his own initiative or in response to requests addressed to him) in a decision-making
process. This widely shared definition shows the inadequacy of the notion
for the questions that have concerned us. The situations that interest us do
not turn so much on available skills and the decisions to be made as on the
modes of organizing the process of production of knowledge (which will be

transformed, but later, into skills that can be called upon) and on the measures to be implemented in order to re-launch the double exploration on the basis of the first lessons. To require a decision to depend on the hearing of experts and counter-experts (representing a wide range of skills and sensibilities), or on the consideration of the points of view of experts chosen from successive increasingly large circles, is to recognize that the inventory of possible positions is complete. It is to consider that the dialogic procedures that permitted the open exploration of these positions can hand over to delegative democracy. Designated spokespersons are able to express identities that have been discussed and consolidated; others, equally legitimate, are in a position to call upon the different, sometimes contradictory results that cooperative research has made available. What is called "expert," in the language of political decision making, generally covers these two categories of spokespersons (those who speak in the name of nature and those who speak in the name of society), who, in the approach we adopt, should be consulted not before the double exploration but after it. Exclusive recourse to expertise turns out to be sterile when hybrid forums are in full activity. On the other hand, recourse to expertise becomes relevant on the return journey, once dialogic procedures have enabled the map of the stakes to be redrawn and these stakes are expressed in a language intelligible to all. Remaining divergences are clearly identified. It remains only to list them and enable them to confront each other through interposed spokespersons and experts.

To talk of expertise (and counter-expertise), and to employ the judicial metaphor to describe the consultation of these experts and the decision making that results from it, is no doubt a step in the right direction.[2] It is to recognize, in fact, that there is a series of mediations between the results of research and their deployment in political decision making. It is also to accept that we do not pass directly from one field to the other. The space between Einstein and Roosevelt, between researchers and politicians, is, as everyone knows, populated by a multitude of experts and spokespersons. Bringing them into broad daylight, organizing their testimony, and getting them to enter into dialogue can only help overcome the serious defects of the double delegation. But if, for cases burdened with serious uncertainties, we were to stop there, we would merely make delegative democracy a little more livable, without opening up the space needed for the development of dialogic democracy. Lifting a corner of the veil that conceals the mysteries of power and revealing the hitherto secret relations between science and politics certainly constitutes progress. But it may also be a formidable weapon against hybrid forums, since, if we stop at that initiative, the possibility of the double exploration itself is compromised. The only purpose of a wide consultation of experts, when decided on *before* the organization of hybrid forums, is to save delegative democracy. It was Jean Bernard, then president of the Comité consultatif national d'éthique, who expressed this logic most frankly and clearly:

When I was consulted, I suggested moreover that apart from theologians, philosophers, jurists, and sociologists, there should also be representatives of the population. The government has only responded very partially to this request by appointing some parliamentarians.... On the other hand, I have always been strenuously opposed, and I am still opposed, to representatives of patients and their families, because the fact of being ill or affected may distort one's judgment.[3]

How could it be put more clearly that the greatest threat comes from ordinary citizens and laypersons, those who are affected by the decisions that will be made and whose judgment is, for that reason, in danger of being distorted! Here Bernard argues for the greatest possible expansion of expertise. The Comité consultatif national d'éthique has no cause to be jealous of Noah's ark. All the passengers are welcomed aboard. Theologians, sociologists, philosophers, ethicists, economists, and parliamentarians are invited to the great embarkation. But if patients present themselves, if citizens suffering in the flesh dare to ask for a place on board, they will quickly be turned away! No doubt it will be very difficult to order this exclusion, but it will have to be made. Experts, theologians, philosophers, sociologists, doctors, biologists, rabbis, and other parliamentarians will no doubt shed a tear, but they will be able to resist the emotion that overcomes them. As experts and legitimate representatives, as wise men, they will make the decisions that serve the general interest, which the persons concerned, blinded by their individual problems, often cannot make out. A plague on demagogy! An expanded circle of expertise and public debate calls for unwavering firmness in making sure that the double delegation is not transgressed. Yes to openness, to a diversity of points of view, and to food and health safety agencies, but on the condition that the frontiers between experts and laypersons, and those between ordinary citizens and their delegates, are guarded with the most extreme firmness.

The reader can see now why we have never put expertise at the center of our reflections. At best, focusing analysis on the problem of the organization of expertise means addressing the question of the return to delegative democracy, after the questions of emergent identities and the conjoint production of knowledge have been resolved; at worst, it means refusing to see that socio-technical controversies are part of situations of uncertainty whose management presupposes that we go beyond the double delegation and its exclusions, for a moment at least.

Returning to the Great Dichotomies?

At no point in this book have we employed the terminology of facts and values or nature and culture. There is nothing surprising about this. These categories go hand in glove with delegative democracy. They are a consequence, or rather a guarantee, of the double delegation. Showing the limits of the latter means prohibiting the use of these words except in some very rare cases where the actors themselves employ them.

Facts and values? The distinction rests on a clear separation between the knowledge produced by scientists and the arbitrary decisions made by politicians. Take the case of nuclear waste. The facts, we are told, speak for themselves: It is objectively proven that the probabilities of contamination linked to deep burial are below a certain level, but nonetheless they are not nil. It is also equally proven that these same probabilities have a higher value in the case of surface storage but that it will be easier to detect possible leakages. These facts being established, and well established, it is up to the political decision maker to make a decision in terms of his values and preferences. For the scientist it is a matter of the description of possible worlds; for the politician it is a matter of the choice of desirable worlds. We have all been trained to consider this distinction legitimate and unbridgeable. Let us acknowledge nonetheless that the frontiers between facts and values are not always as clear as is claimed.

We can all cite without difficulty ten examples of decisions made under circumstances of great confusion. How many scientists and engineers have not benefited from the incompetence or benevolence of their political interlocutors to take them in and deceive them into thinking there is only one technically possible decision, the one on which they have been working? ("My dear Minister, the fast breeder nuclear reactor is the only realistic option, both economically and politically!" "Prions are responsible for BSE, Mr. Civil Servant, but they will never be able to cross the species barrier! The stubborn facts are there and it is useless to ignore them!") Conversely, who cannot call to mind the words of decision makers who, playing on the disagreements between experts, harden the point of view of some of them in order to justify the decision that settles them politically? ("Researchers assure us that the planet is getting hotter and that traffic is the main cause of this, so we will tax road transport!")

In view of what appears to be a confusion of genres, two attitudes are possible.

The first attitude, which aims to save the distinction between facts and values, persists in asserting that if we allow ourselves the means—that is, if we maintain even more strictly the cut between science and politics—the confusion can be avoided. Unfortunately, and even those who want to maintain the divide between facts and values acknowledge it, no objective fact established in a laboratory can be invoked as it stands as an absolute necessity, an inescapable constraint; the concrete reality of the world in which we live is too complex for simple transpositions to be possible. This argument, which we have ourselves used to establish the space of hybrid forums, is deployed here not to emphasize the importance of research in the wild but to justify recourse to an organized multiple expertise. Here again the solution consists in saving the double delegation, and the split between science and politics it implies: absolute certainties about the real and complex world do not exist, but the organized and transparent confrontation between experts leads to reasoned evaluations that enable us to see things more clearly in the light of the current state of knowledge. The judicial model is a good model. First the experts give their point of view; then the judge (here the politician) decides in all conscience and honesty. The translation from the laboratory to the outside world is organized in such a way that delegative democracy is not threatened. The Prince is advised by a host of experts with the widest possible range of competencies and backgrounds.

The second attitude, the one we endorse, consists in taking equally seriously the existence of an irreducible distance between the facts established by secluded research and the problems encountered by laypersons and ordinary citizens. But instead of trying to reduce this gap at all costs by organizing a wide and open consultation of experts, the challenge is accepted of introducing research in the wild into the game while favoring the formation of new identities. Instead of resorting to established experts and the maintenance of the monopoly of secluded research in the investigation of possible worlds, and instead of resorting to institutional spokespersons who keep emergent concerned groups at arm's length, the setting up and organization of hybrid forums is favored. Now the distinction between facts and values is not only blurred in these forums, it is quite simply suppressed. It is through the exploration of possible worlds that identities are reconfigured, these identities leading in turn to new questions. For example, by closely linking together the two explorations, those suffering from myopathies advance knowledge at the same time as they further the recognition of an identity fashioned by genetics. At this point facts and values are so interlinked that the distinction between them is no longer pertinent;

subjectivity lives on an objectivity that it questions and problematizes in turn. This spiral is made possible by procedures that organize the to-and-fro movement between the two axes. Conversely, the consultation of experts, when it precludes hybrid forums, aims to maintain the separation between the two axes and make the transition from delegatory to dialogic democracy impossible. This is why the return to the distinction between facts and values should only be envisaged once the double exploration had been completed, when the question of identities and possible worlds has been clarified. Facts and values really exist then, but instead of being constituted as starting points, they are seen as an outcome. Hybrid forums are the crucibles in which existing facts and values are mixed in order to be recomposed and reconfigured.

Nature and culture? This distinction is also behind us, or rather it is simply a possible result and not a starting point. As Bruno Latour has shown, it takes up the distinction between facts and values in a more general mode, and like that distinction, it paralyzes political debate.[4] It does not merit lengthy treatment here. No actor in the world refers to it any longer, except in passing, in order to found his or her claims, or to justify procedures or institutions. It is not by chance that all the experiments conducted within hybrid forums are carried out in the name of two requirements: to enable ordinary citizens to have their say and to break the monopoly of the specialists. What do those suffering from myopathy, those affected by AIDS, and the farmers of Bure say? They certainly do not say that they are struggling to redefine the frontiers between nature and culture, but quite simply that they are struggling for a reorganization of the political debate and of expertise. In keeping the notions of nature and culture alive, specialists of the social and human sciences contribute to an enterprise of concealment whose only result could well be to save the double delegation by acting as if hybrid forums did not exist.

Let's listen once again to those with myopathy. Is their identity natural, or cultural? Neither one nor the other, but both at once. It is constituted from genes that have been socialized by the community that those with myopathy strive to compose. They are "clothed," civilized genes. Hybrid forums subject nature and culture to the same treatment as facts and values: they brew them, mix them to the point of making their distinction non-pertinent, indeed dangerous. The actors ask themselves concrete questions: How can we organize cooperation between secluded research and research in the wild? How can we allow emergent identities the possibility of constructing their identity? It would be out of place for the social sciences to be deaf to these questionings and to make them disappear by "raising the level" of the debate and imposing metaphysical concerns on the actors—concerns these same actors go to great lengths to ignore so as to pass from delegative to dialogic democracy. The distinction that should be favored is between that which it has been decided to investigate or reshape and that which it has been decided not to question.

Use the Mechanisms of the Market?

At a time when everybody swears by it, it might seem problematic simply to say nothing about the economic market and its iron laws. The question has not yet arisen. However, it is no longer possible to back away. Is it in fact reasonable or realistic to write dozens and dozens of pages, use a lot of breath, and demonstrate our eloquence when all the cases examined— GMO, nuclear power, the invention of drugs for treating rare diseases—are economic cases? The resolution of the problems posed is, above all, a question of the efficient allocation of scarce resources. Serious people tell us "Come down to earth! Yes, everything is political. Certainly democracy must be democratized. But the best way to achieve this is to organize genuine markets which enable different agents to make a realistic choice of the most efficient solutions." Should we not take them seriously? "By dint of trying to enrich democracy," they continue, with the hint of a smile on their lips, "you will end up impoverishing citizens and consumers. Stop gesticulating! Political lyricism is one thing, economic realism is another. Certainly GMO and waste are important subjects, but if we spend all our time discussing them, before long it will all be so costly that there will no longer be any waste to incinerate and not a bean to eat. The best saturnalia are those that last only a short time. Leave these forums, even the hybrid ones, and get back to work!"

Such disquiet is understandable. It is easy to show that it is excessive. The fall of the Berlin Wall and the collapse of the so-called planned economies have opened the way to a serene reflection on the question of markets and their management. Yes, markets are tools whose efficiency in the production of wealth and well-being is unequaled to this day. But they must be organized for their social yield to be optimal, and their organization must be the object of thorough reflection. Nothing is worse than a market abandoned to itself, for it quickly ends up producing irreparable damage. A market is a high-precision machine that presupposes constant tuning, impeccable maintenance, and attentive after-sale service. Economists saw this, and long ago they introduced the idea that under certain conditions the market could have serious weaknesses. One of these is central for our purposes. It is captured in one word: 'externality'.

Economic externalities are a particular case of what we have called *over-flows*. What is an externality? The simplest thing to do here is give an example. Consider a chemical factory that produces aluminum and emits chlorinated fumes. The spread of these fumes threatens the cattle rearing and agricultural cultivation of neighboring farms. To combat or eliminate the disagreeable or harmful consequences of these fumes (animals losing weight, low crop yields), the farmers affected are forced to make investments. Now (this is where the notion of externality becomes pertinent), without incentives, the firm concerned does not include in its calculations the cost that it passes on to agents (farmers) who, although penalized by its activities, remain outside its sphere of economic relations. In doing this the firm produces externalities, which in this case are negative: the farmers' interests are affected without them being able to defend them or get them taken into consideration. In fact, if they decide to continue their activities, they must make investments for which they receive no compensation. To get others who have no say in the matter to bear part of the cost of one's own activities is a common practice for economic agents, but one that compromises the efficiency of markets: it is easy to show that there is not an optimal allocation of resources.

We can now introduce the general definition of externality. Consider three agents (A, B, and C) who are involved in a market transaction or, more generally, in the negotiation of a contract. In the course of the transaction or negotiation of the contract, these agents express their preferences or their interests and proceed to evaluate the different possible decisions. The decision made has positive or negative effects, which we will call externalities, on three other agents (X, Y, and Z), who do not play a part in the transaction or negotiation, either because they do not have the means to take part or because they do not wish to. Not including in one's accounting the effects produced by one's activities on other agents is the origin of the overflow called *externality*.

The reader will, without difficulty, make the connection between the existence of hybrid forums and the production of externalities by existing markets. All the examples we have given, from the first chapter of this book, enter at least in part into the category of externalities. Nuclear waste? The problem it raises is a consequence of the fact that when France committed itself to nuclear power the points of view and interests of the future residents of burial or storage sites were not considered. And it was not only their point of view that was not considered, but also that of future generations. The definition we have given of 'externality' has the merit of accommodating those agents who do not yet exist. X, Y, and Z may be our children, our grandchildren, or our great-great grandchildren, who cannot take part in the calculation of decisions but who will nevertheless have to deal with their consequences. People with myopathy? Here is a group that is ignored by the major pharmaceutical firms, for its members, far too few for their demand for drugs to be thought profitable, are struck by what is rightly described as an orphan disease. The economic agents do not integrate their demands and interests in the decisions they make. They are quite simply outside the field of vision of the firms and, as a consequence, that of the research laboratories; they are excluded from the market, just like the future generations we have just mentioned. People with AIDS? Here again the market, with its myopia, is blind to a part of their concerns; what interest do pharmaceutical laboratories have in passing directly to triple therapy when the investments made for single or double therapies have not yet been recuperated? BSE? Why should enterprises producing animal feed be concerned about the distant and uncertain consequences for the consumer of beef of recalcitrant prions that will turn out to be sufficiently supple to cross the species barrier? GMO? How can we expect Monsanto or Novartis to include in their calculations the effects of the possible dissemination of resistant genes, or to take account of the consequences of the generalization of transgenic plants for North-South relations or for agriculture? The market is efficient because it is able to frame the problems and not get entangled in all the overflows and side effects that it might generate. It would be ridiculous, indeed counterproductive, to ask multinational corporations to concern themselves with their overflows before they have taken place, for they would be paralyzed.

Markets, when calculating interest, profits, and returns on investments, draw a strict dividing line between that which is taken into account and that which is not. This is where their strength lies, since they can be deaf to the protests of residents, spokespersons of future generations, or orphan patients. But it is also what marks their limits. This frame, with the exclusions it generates and the overflows it produces or tolerates, is at the origin of matters of concern and of the issues which cannot be dissociated from the concerned (emergent) groups who express them and make them visible and debatable. Thus, the market, left to itself, tends to produce injustice, forgetting and ill-treating our descendants in one case and suffering minorities in the other. Just as it is legitimate not to saddle enterprises with the still-virtual burden of the overflows they create, so it would be idiotic to prevent all the groups who feel that they are possible victims of these overflows, or that they are affected by them, from making their voices heard and giving public expression to their concerns. Hybrid forums are not the simple consequence of the limits of delegative democracy. They are set up on terrain left fallow by economic markets; they mobilize actors who reckon that their identities, problems, and concerns are not taken into account by those whose accounting decides the allocation of scarce resources. Markets, in alliance with the technosciences, are thus constant sources of (as yet) unstabilized matters of concern. As their ascendancy increases, the need for dialogic democracy intensifies. Once the explorations are (provisionally) complete, and identities (temporarily) set, market arrangements can be designed to allocate resources and to coordinate supply and demand.

Markets and delegative democracy work hand in glove, moreover. They mutually reinforce each other. Both presuppose framings that avoid constant overflows. Just like delegative democracy, the market has a horror of deep uncertainties. That is why it too relies on secluded research. Letting open cooperation between secluded research and research in the wild take hold for too long is out of the question.[5] So too there is no question of its accompanying emergent identities in their first tentative steps. It is, of course, on the lookout for new needs that it can express, but it has every interest in waiting for these identities to become consolidated and creditworthy. Hybrid forums are therefore as useful to democracy as they are to the market economy. They organize the identification and exploration of externalities and exclusions. They also measure externalities and exclusion, so that these can, as economists say in their somewhat barbaric vocabulary, be internalized—recorded and included in the calculation of costs. At the end of the double exploration of possible worlds and the collective, it is possible to draw up the balance sheet of externalities and exclusions, measure them, and then take measures for taking them into account. Without the hybrid forums of nuclear waste, GMO, BSE, or orphan diseases, it would be impossible to draw the map of the overflows and exclusions and to say to economic agents "The uncertainty is over, here are the proven effects of your activities, here are the groups concerned, and here is the price they attach to that which may put an end to these overflows." Then, but only then, does it become possible to reorganize the markets (which means "internalize the externalities") and to see to it that, after debate and negotiation, the firms producing aluminum take responsibility for a part of the costs they induce, or that the enterprises that propose sowing GMO take into consideration the consequences of this new technology for the developing countries or for the organization of agriculture, or that pharmaceutical firms contribute to therapeutic research on orphan diseases.

In a situation of uncertainty, calculation and negotiation of interests are impossible without the double exploration that duly organized hybrid forums permit. The market, a formidable tool for arriving at compromise between established and contradictory interests battling with each other, can be reorganized only after this investigation is completed. In the absence of hybrid forums that extend, debate, and reorganize them, markets quickly become contested, illegitimate, and sources of inequity and injustice.

Liberalizing markets does not, as some would have us believe, mean putting them beyond discussion; rather, it means favoring the expression of every voice, facilitating the expression of views that, organized in hybrid forums, enable the effectiveness and social legitimacy of existing markets to be enhanced by working at their reorganization. The issue is not who is for or against the market, or whether there should be more or less market. The political question concerns the forms of organization of markets.[6] Let us free markets from the supposedly natural laws that the most extreme liberal doctrines attribute to them, so that they are able to take in the proposals produced by the hybrid forums that manage their weaknesses.

Whether it is a question of the calculation of risks, of quarrels about the separation of facts and values, of nature and culture, or of the laws of the market, the conclusion is the same. All these categories, when they are invoked, imprison the protagonists in the iron cage of delegative democracy, reinforcing the fixed character of the double delegation. Consequently, they prevent us from thinking about the symmetry between secluded research and research in the wild, just as they prevent the constitution of emergent minorities being taken into consideration. They deny the fecundity of hybrid forums, which nonetheless enrich democracy and free markets.

So we have rid ourselves of the great narratives that conceal the concrete experiments in which actors have been involved for several decades and through which they strive to find solutions to the practical questions that they raise. We can now size up the full extent of the procedural innovations developed in the hybrid forums, or at least some of them, in order to examine in what respect they could contribute a general and satisfactory solution to the insistent question of the representation of minorities.

A Procedural Innovation of Technical Democracy: The Representation of Minorities

Democracy constitutes both a fixed horizon and a never-completed undertaking. This double acknowledgement is expressed in the collective statement "democratization of democracy."

The desire for more democracy applies notably to the question of representation. How can we be sure that, at the time of the composition of the common world, everyone has been able to make his or her voice heard, and that this voice has been taken into consideration? There is no general answer to this question. What we call representative democracy is an institutional bricolage that differs from one country to the next, depending on historical trajectory. It is an assemblage of procedures in multiple forms, resulting from several centuries of ongoing experimentation and cross-fertilization. We have to acknowledge that this bricolage currently comes up against the insistent question of the mode of representation of minorities.

Consider the very pragmatic definition of representative democracy proposed by Christian Delacampagne. At first sight, it seems to do justice to the various and sometimes contradictory definitions that are usually advanced. "Representative democracy," Delacampagne writes, "is in principle a parliamentary democracy: parliaments are assemblies of men and women, more often men, chosen for their wisdom and whose deliberations are supposed to arrive at the best possible decision." "But," he notes, "if the existence of parliaments is necessary, it is not sufficient." And respect for three principles should be added:

The *principle of tolerance* that requires the state to assure on its soil the free expression of beliefs, and political, philosophical, or religious ideas, provided that the latter do not cause harm to public order. The second principle is that of *the separation of powers* whose objective is to establish the *rule of law*, that is to say, to protect the citizen from any abuse, and in particular, obviously, from the arbitrary use that those with public authority might be tempted to make of this authority. The third principle is the *principle of justice*: a democracy worthy of the name must not be satisfied with being a formal democracy, blind to the inequalities separating some from others, it must set its sights on a concrete end of social justice.[7]

This definition has the advantage of emphasizing that representative democracy is never an established fact, for its implementation passes through procedures that, like any procedure, often end up producing results opposed to those for which they were devised. But one of its obvious limits stems from the place it gives and the role it allots to the notion, or rather the principle, of tolerance. It is one thing to ensure the free expression of what Delacampagne calls "beliefs, and political, philosophical, or religious ideas"; it is another thing to take them into consideration at the time of the construction of the public order. The principle of tolerance, as defined by Delacampagne, inspires procedures intended to set out a space for the minorities who want it. Like travelers, they are guaranteed the right to set themselves up in the vague areas that no one claims, under the express reservation that they are not to disturb the peace of neighboring residents. But the principle is of no use when it is a question of resolving the concrete question of the participation of these same minorities in the work of the design and composition of the collective to which they will be parties. The notion of tolerance, however generous (or perhaps because of its generosity), is dangerous, for it transforms a profoundly political question into a simple problem of coexistence. It allows one to think that the question is settled when in fact it is awaiting settlement. Hence the need to continue with the work of institutional bricolage, of the enrichment of procedures, in order to move from tolerance ("We support you on the condition that you be good students") to involvement ("You are qualified not only to express your point of view and to defend it, but also to take part in the search for a common world").

This requirement is all the more pressing because the representation of minorities is one of the thorniest problems for advanced democracies. What kind of procedures could enable minorities to be actively associated with the composition of the collective? In other words: How can we reconcile, on the one hand, rights linked to groups that define themselves by their own, specific identity, demands, and forms of solidarity, to which they are attached above all else, with, on the other hand, the organization of a common world that, in one way or another, presupposes compromises and renunciations, since the simple assertion of the rights of some may come into conflict with the rights of others? What place is to be accorded to minorities in the collective? What weight is to be given to their demands and their interests? How can their particular wills be taken into account in the expression of the general will?

We would like to suggest that the procedural innovations devised by hybrid forums could serve as models or at least sources of inspiration. They constitute, in fact, an irreplaceable laboratory in which representative democracies learn how to deal with minorities. In this book we have constantly encountered and rubbed shoulders with minorities: people with myopathy, AIDS patients, residents of sites for the storage of nuclear waste, people living near a chemical factory or a discharge of toxic products, consumers of transgenic food. These minorities, like all minorities, whether ethnic or religious, battle to be recognized and heard, and mobilize to be represented. They remind us that, in every area, representation is a permanent and open question for democracy.

Hybrid forums shed new light on the confrontation between identities. By displaying emergent minorities struggling not only to find their own identity but also to have it recognized, they show us that the trajectory followed by identities is more important than the identities themselves, when the goal is to construct a common world. Hybrid forums also reveal that these transformations of identities, without which the construction of a common world soon becomes a utopia, are inseparable from the investigations, explorations, and scientific and lay experiments enabling groups to reformulate their problems along with that which they value and that which they are prepared to forgo. In this way possible worlds proliferate, and the arrangements necessary for the construction of a common world become easier.

At this point we need to mention John Dewey and his definition of the public. Dewey contends that, confronted with unexpected overflows, the state (the other name for delegative democracy) is powerless. It constantly has to be reinvented, shifted and taken in charge by a multitude of different, fragmented publics affected by these diverse overflows. These publics launch inquiries to explore the issues and the evolving and changing networks connecting them. With Dewey the unsolved tension that hybrid forums create between delegative democracy and dialogic democracy has its counterpart in the constant re-creation of the state by the emergent concerned publics. His solution contrasts with that of Walter Lippman and the revamping that he proposes for delegative democracy. With Lippman the public is called upon when the experts and decision makers, overwhelmed by the complexity of the problems confronting them, have no other options available. The public thus saves delegative democracy which, outside of these dramatic episodes, wants it to be apathetic. Lippman likes the public most when it is a ghost, non-existent, and only episodically grants it an active existence—in the form of nothing more than a rescuer! The creative and open dynamic described by Dewey contrasts with the managerial, closed logic of crisis management imagined by Lippman. It is compatible with that of hybrid forums and their way of composing identities, by granting emergent minorities the most attention and a key role. Dewey nevertheless says little about the procedures enabling the publics, necessarily in a situation of weakness, to play their part and especially not to be swallowed up by a powerful state apparatus. The tolerance referred to above would hardly be enough to save those who still don't count. It is on this point—the conception and implementation of the procedures making a dynamic of composition of the collective possible and necessary—that reflection on hybrid forums can be of general relevance.[8]

When the question is the chador or clitorectomy or even Ireland, there is indeed a strong temptation to speak in generalities and to appeal to the great principles, instead of sticking to a reflection focusing modestly on procedures. The protagonists are referred to the general will, national sovereignty, or Reason, as so many threats intended to silence those who dare to speak of their particular problems. This is because defense of wearing the chador, like the claims of Irish identity, seem to be rooted in traditions that are firmly established and therefore difficult to negotiate. In these cases, the composition of the collective through successive adjustments represents a huge challenge. Imagination is often lacking when it is a question of devising procedures.

But it is easier to halt this propensity for confrontation when it is a matter of simple peasants opposed to the installation of a site for nuclear waste, or patients demanding the establishment of compassionate protocols. In fact, identities are emergent in hybrid forums. New, unforeseeable groups emerge, take shape, and are transformed, their still-inchoate existence being created by decisions or activities that may themselves be revised. Their interests are malleable, their demands are open to debate. Not only is there no firmly established and constraining tradition to be invoked, but in addition the problems appear to be contingent and their resolution does not seem to be insurmountable. In hybrid forums, minorities raise questions whose answers can be found without too much difficulty, on condition that there is agreement to do everything possible to find them. That is why the socio-technical controversies that we have examined until now are remarkable laboratories for refining and testing procedures whose generalization to less emergent situations could then be considered. They should enable us to advance in the art of politically and democratically managing the difficult question of minorities.

The hypothesis of the exemplary character of hybrid forums and the possible generalization of the lessons that may be drawn from them, notably as regards the management of minorities, nevertheless poses a formidable problem: Do the transposition of procedures, the rules of consultation, and the rules for the organization of discussions suffice to resolve, as if by the wave of a magic wand, the question of representation? Would it not be better to return to the foundations of democracy, in order to give more thought to the meaning of founding notions like those of the general will or citizenship? In allowing ourselves to be seduced by the innovations of technical democracy, are we not in danger of abandoning the terrain of political reflection for the more reassuring but less fruitful terrain of rules and procedures? Are we not replacing the nobility of the debate of ideas with discussions lacking grandeur on the functioning of organizations? Will not the enchanting world of political action not collapse into the grey and tedious world of social engineering?

If the answer to these questions were positive, it would be pointless to seek to transpose and extend the procedural innovations developed by hybrid forums. For these procedures to be generalizable, it would be necessary that the equitable character of the decisions was not due to the decisions themselves but rather was conferred on them by the procedure. This is what we shall endeavor to suggest now by examining successively two closely connected questions. The first concerns the reasons why certain decision-making procedures are equitable, that is to say, produce in those that they concern the intimate conviction that the decisions made are their decisions and that they are legitimate and good. The second question concerns this strange transfer of qualities: How can we explain the phenomenon that procedures that are judged to be equitable produce equitable decisions?

Having established the validity of procedural justice, we will examine some objections to it. Does not the precedence accorded to procedures favor the manipulations and Machiavellianism of those who are familiar with its labyrinths and twists and turns? Is it not in contradiction with the requirement of efficiency? These two questions will have to be answered if we are to give plausibility to the transposition and generalization of procedural innovations produced by technical democracy in action.

Equity or Feeling of Equity? The "Fair Effect Process"

Under what conditions can the way of making a decision influence how actors evaluate its equity? Or, to use an expression proposed by some psychosociologists,[9] how is the "fair effect process" to be explained? This expression designates the mechanism by which actors become convinced that a decision is equitable. This mechanism, which falls within the province of social psychology, has been the object of empirical studies that have enabled its workings to be identified.[10]

The "fair effect process" is linked to the degree of control exercised by actors over the process of the development of the measures that will be taken. Concretely, what is at stake is control of what political scientists call the third party and which, in delegative democracy, is embodied in the public authorities and the state apparatus. That each group has been able to express its point of view, can observe that its point of view has been taken into consideration and discussed, and, *in fine*, that the measures taken have been decided impartially, is one of the elements that, when present, produce the conviction in the actors concerned that the decision is just. This requirement disqualifies in advance decisions concocted in the dark offices of public bureaucrats.

Decisions are also judged to be more equitable the more the relations established between those who are party to it are stamped with trust. Classically, trust is defined by the fact that, to undertake an action, agent A leaves it up to whatever agent B says, promises, or does without seeking to verify or check himself what B says, promises, or does. This definition, which is applied to relations between two agents, can be extended without problem to relations of trust between an agent A and an institution B (like science or a banking organization). Now trust is a mental disposition produced by appropriate procedures. Trust is generated notably by the existence of impersonal arrangements of certification (which confirm and assure that certain predefined rules have been followed and which sanction deviant behavior) or even by the repetition of interactions (in this case, the agents honor their commitments so as to avoid losing their reputation). This is why formalized procedures clearly setting out the rules of consultation, prescribing the mechanisms of the expression and consideration of points of view, and keeping track of the deliberations leading to the final decision, help to create this climate of trust which favors the "fair effect process."

The same (sociological) studies have also highlighted the fact that when the "fair effect process" produced by the given procedures exists for the groups concerned by the decision, then it is naturally extended to groups not concerned and possibly not involved in the deliberation or consultation. If the farmers in the vicinity of a nuclear waste storage site fighting to have the issue of the burial of nuclear waste re-opened, or the AIDS patients who take part in a drug trial, reckon that the decisions concerning *them* are just, then all those who are not directly involved will tend to share that evaluation.

Equity of Procedures, or Equity of Decisions?

It is one thing to show and acknowledge that the feeling of equity is generated by the procedures implemented, but it is something else to view the decisions made in this way as *intrinsically* equitable. Should we not take seriously the classic objection that it is not because a decision is judged equitable that it really is so? Laypersons, however diverse and informed they may be, are not necessarily in the best position to judge dispassionately on justice.

For those who dispute the idea that the feeling of equity is the same as equity, there are many reasons for this blindness. The first is the supposed myopia of ordinary citizens and non-specialists. Owing to their conditions of existence, they find themselves confronted by short-term problems and have neither the resources nor the motivation to project themselves into the long term.

Let us acknowledge that this preference for the short term exists, but let us add that it could not be a general rule. Would it be fair to say that those with myopathy think only of their present or immediate interests when they provide an exemplary and striking illustration of long-term commitments which notably find expression in considerable investment in the most basic research? If there is a preference for the short term, then it is manifestly found on the part of the public authorities, firms, and, as an indirect consequence, academic research laboratories! Can we assert without batting an eyelid that groups who discuss the options for managing nuclear waste are not interested in future generations, when it was precisely some of these groups who helped introduce this strange category into the debate? That they did so with ulterior motives, so as to defend their immediate interests better, hardly matters. That egoistical concerns are concealed behind anxieties about the future does not prevent the latter from being taken into consideration. What matters is that the long term is discussed. The list of examples could be extended, and we would find few cases in wide-open forums where any group of laypersons would not defend the long term. The more dialogic the procedures, the more likely it is that the future will be explicitly taken into consideration for the double exploration of possible worlds and collectives.

The second possible reason given for the blindness that would be the source of a discrepancy between actors' evaluations and the reality of the measures taken is the fickleness of their judgments. A group that favors a particular decision one day will be vehemently opposed to it the next. And how can we talk of equity when every position can be reviewed, when the interested let themselves be swayed by the opinion makers or, alternatively, let themselves be pressurized by current events, and when that which seemed to be just one day is in danger of being seen as iniquitous two days later? This objection is well founded. It is even completely relevant. In situations of uncertainty, of emergence, no preference is stable; no criterion of judgment is firmly established. Hybrid forums are, in essence, apparatuses that generate turnarounds in opinions and encourage the review of the best-established agreements. But it is precisely so as to handle these situations of uncertainty, to the extent of possibly stabilizing them, that the procedures we have presented were devised and tried out. It is in order not to remain powerless in the face of these uncertainties that it was necessary to abandon the reassuring idea of clear-cut and definitive decisions. As we can see, the objection does not apply to hybrid forums but to delegative democracy, which has no means for taking these increasingly frequent instabilities into account. The challenge raised by technical democracy is how to devise procedures that can take charge of these instabilities.

The third reason often advanced to refute the idea that equitable procedures produce equitable decisions is symmetrical to the previous reason. It emphasizes the definitive incompatibility of points of view and interests. How can we imagine any kind of justice when expectations are incommensurable and when there is no space in which equivalences can be postulated? Whether points of view are unstable or, on the contrary, seen to be rigid and non-negotiable, in both cases the very idea of equitable compromise becomes unrealistic. This is an excessively pessimistic view. The search for a common world in hybrid forums is made possible by the unstable character of identities, the flexibility of positions and representations, and the malleability of knowledge. These different identities can be brought together and adjusted to each other precisely because they are unstable and can be transformed. The main property of the material on which hybrid forums work is its capacity to be fashioned. It is neither definitively volatile nor definitively rigid.

The usual objections miss their target. Equitable procedures are precisely procedures that are designed to facilitate the expression and consideration of the greatest possible diversity of points of view and sensibilities; they are more able than others to bring together positions involving the long term. Equitable procedures are also procedures that allow identities the space they need in order to emerge, be transformed, and be composed with each other, notably by leaving the collective exploration of possible worlds open. Undoubtedly, the argument stands up to the criticisms. An equitable measure is a measure taken by following procedures that produce in all the protagonists the conviction that it is equitable! And, as the reader can easily verify, the dialogic procedures of hybrid forums are equitable procedures. (See box 7.1.)

Recognition that an equitable decision is one made in a way that is judged equitable and that this necessarily negotiated judgment is produced by procedures that will be described as equitable inasmuch as they produce this particular effect amounts to a decisive advance. But before taking the

Box 7.1

Are dialogic procedures equitable?

To conclude on the exemplarity of hybrid forums it is necessary to confirm that dialogic procedures are equitable procedures. In other words, we must show the compatibility between the criteria presented in chapter 5 and those we have just enumerated: Do procedures with a high degree of dialogism assure the involvement, control, and trust of the concerned groups?

Consideration of the points of view of the different actors?

This property of the procedures corresponds to several criteria of table 5.1. The earliness of layperson involvement, the diversity and independence of the groups consulted, the representativity of the spokespersons, as well as the assurance of equality are all variables which give a precise meaning to the general notion of consideration. Similarly, the capacity of procedures to generate in the protagonists a constant concern for the collective helps to facilitate the consideration of their singular points of view, no third party having the de facto monopoly of the definition of the general interest.

Capacity of the actors to confirm that their points of view have been considered in the process of working out the measures to be taken?

The criterion of traceability and transparency is an answer to this question. So too is the more general criterion of the organization of the public space of debates, which enables us to give a richer meaning to the simple constraint of the integration of discussions within the political decision-making process.

Establishment of relationships of trust?

This requirement is also found in the table of criteria. When we evoke the seriousness and continuity of the expression of points of view, but also the clarity of the rules of organization of debate as well as the need for traceability, all we are doing is defining the terms of reference to which the procedures must correspond in order to ensure trust between actors. Continuity entails the repetition of interactions, a repetition that is at the root of effects of reputation and assures confident involvement. Traceability is the equivalent of a mark of quality which also guarantees trust in commitments, but this time in an impersonal manner.

reduction of substantive justice to procedural justice to be an established fact, we must answer two objections that are often made to this pragmatic approach—two objections concerning manipulation and efficiency.

The Risk of Machiavellian Manipulation

Procedures are often suspect. For some, they deter us from getting to the bottom of things. They slow down debate, divert it, and lead it astray. They come between the actors, and as a result they are a choice terrain for all those who have the resources that enable them to manipulate procedures in order to achieve their objectives and defend their interests.

Are not procedures traps laid by decision makers who, by limiting debate to formal questions, avoid debating the decisions themselves? "Let's talk about procedures, if you really want to! Let's talk about nothing else, we will waste our time!" Is not the decision maker's supreme cleverness his ability to manipulate procedures with a view to producing the "fair effect process" and so the feeling of equity that will enable him to get measures accepted that are iniquitous but taken in the correct way? Moreover, it is because these dangers exist that the notion of participatory democracy has a bad press and we have refrained from employing it.

These fears are widely exaggerated. What is essential for ordinary citizens and laypersons in dialogic democracy is not participating, but weighing up and contributing. In chapter 5 we highlighted this point by showing the importance of the construction of a public space that enables dissident voices to be audible and emergent identities to be perceptible. The possibility of manipulation and the skills of professional rhetoricians are limited when procedures are clearly and rigorously defined, and when they are made constraining and debatable. They are anything but simple instruments that can be controlled by a wily politician in pursuit of his own objectives.

Let's consider the example of the "Bataille law." This law, conceded by the French parliament to all those opposed to deep underground burial of nuclear waste, made the continuation of research on other technical options, surface storage, and transmutation mandatory. When it was passed, many observers cried manipulation. They accused the government of having pretended to open the range of choices, only to revert to the underground burial option 15 years later when its initial commitments had been forgotten. We now know its effects. Whatever the legislator's intention, and whatever the Machiavellian artfulness of those who supported it, it ended up completely reorganizing the political game, imposing new identities and the consideration of previously inaudible demands. The history of nuclear power should thus be entirely rewritten, bringing calculating actors on to the scene, but calculating actors who have gradually been constrained to apply their calculating virtuosity, not in the decisions themselves, but in devising the procedures adopted for making these decisions.[11]

A dialogic procedure is a promise to be kept, an invitation to broaden and deepen the debate; it brings, inscribed in it, the possibilities of circumventing the political elites and experts. Criticisms directed at a particular procedure, and especially at its susceptibility to manipulation, are an encouragement to make the procedure more dialogic by imposing more strictly the terms of reference set out in chapter 5.

To play down the influence of the actors' cynicism and tactical skill is not to sin by excessively naive optimism. Yes, the actors are calculators, cynics, and Machiavellian. But tactical skill is not the monopoly of any one group of actors; all of them possess it, and they are equal when this skill is applied to the calculation of procedures rather than decisions, and when this calculation takes place in a public space open to groups dissatisfied with existing procedures. The design of procedures actually requires less professional expertise than the elaboration of lengthy argument on the probability of prions crossing species barriers.

The advantage of focusing debate on the content and conditions of implementation of procedures is to bracket off ulterior motives by leaving them out of the public debate. Everyone has the right to nourish the most perverse and anti-democratic motives imaginable. All that matters is the procedure of consultation and working out the measures to be taken. Partisans of nuclear power doubtless seek to defend and develop it by skillfully taking advantage of public debate. However, the nuclear technology resulting from these debates will be completely different socially, politically, and even technically from that of a form of nuclear power decided outside hybrid forums. To speak of nuclear technology in general has no meaning. To play the game of for and against is even more inept. The CEA engineers who invent new options and new configurations for storing nuclear waste are well aware of this. If the return to procedural forms produces observable effects even in the case of French nuclear technology, where there has been an accumulation of unilateral decisions in order to produce irreversibility, then these effects can only be more visible and significant in all the other cases.

Every procedure generates overflows, every objectivized procedure lends itself to manipulation, but the continual evaluation and transformation of these procedures increases the cost of manipulating them. And at any time existing procedures can be made more dialogic or, if necessary, new ones can be invented.

Efficient Because Dialogic

Although we have advanced in the demonstration of our thesis, we have
still not finished with the objections. Until now we have focused our attention on the crucial question for democracy of equity. But with regard to decision making, this criterion is not enough on its own. Decisions should,
certainly, be as just as is possible; that is, they should take into consideration the points of view of each in a way which is judged to be equitable.
But we also need to demonstrate their efficiency. And on this topic also,
those who plead in favor of maintaining the double delegation do not lack
arguments. In cases of high uncertainty, which correspond precisely to
hybrid forums and the procedures we are discussing, the adversaries of
dialogic procedures emphasize that the resolution of problems calls for a
significant increase in the cognitive resources called upon, and not for
greater openness in the consultation of points of view. What is to be
sought is even more expertise, even more diversification of expertise, even
more rigor in the organization of the work of experts, and not more impli-
cation and involvement of laypersons. To arrive at rational measures, we
must know how to resist demagogy, as tempting as it may be, and the risks
it runs of getting the issues bogged down!

The reader will have understood that this objection does not stand up.
It artificially opposes experts and laypersons, legitimate representatives
and ordinary citizens. Dialogic procedures are not intended to eliminate
experts. They are intended to organize cooperative research between specialists and concerned groups: this collaboration will be all the more fruitful as the experts involved master diversified skills. Dialogic procedures do
not exist in order to eliminate representation. We have emphasized that
without representation there would be no democracy, and that no conception is more mistaken than that which maintains the illusion of a pure and
simple transfer of will from the represented to his or her representative.
Quite the opposite is the case. In hybrid forums the spokespersons' legitimacy and pertinence are obtained thanks to constant interactions with
those for whom they speak and with whom they take part in working out
what is to be said. Owing to the complexity of the information to be produced and taken into consideration in hybrid forums, and owing to the
emergent character of the identities, no elite—however enlightened, diversified, and rich in multiple competences, and however well equipped—can
cover all considerations. It will always lack some of the information necessary for the complete formulation of the problems, and it will very quickly
find itself making decisions that will be rejected and judged unacceptable.
The logic of dialogic procedures is to organize consultation *and* the devel-

opment of measures in such a way that *both* the complexity of the questions *and* the wealth of answers are preserved. Dialogic procedures make easier a sort of collective intelligence. Not directed against the experts and representatives, they are intended to re-immerse experts and citizens' representatives in a milieu and a dynamic from which they tend to cut themselves off and which nevertheless can only enrich the measures taken and make them more reasonable.

Dialogic procedures do not stage the often-announced battle between meritocracy and democracy; they do not choose the people against the technicality of mathematical theorems or the laws of physics. The odds are that they give every guarantee to specialists that they can work in the best conditions, within their laboratories, surrounded and valued by their peers, while encouraging them not to cut themselves off from the world. Concerned groups are not interested in the technical content of this or that piece of research if no connections have been made between their concerns and those of the specialists. Giving prominence to these connections, discussing their reality, and elucidating their nature are legitimate matters for discussion in hybrid forums. The staunch partisans of the double delegation can be reassured that no one will put the demonstration of Fermat's theorem put forward by Wiles to a vote! No one will propose the establishment of a hybrid forum on Hilbert's nth conjecture! On the other hand, it may be that in several decades the succession of hybrid forums will end up having an effect on the problems posed by mathematicians. After all, the myths say, geometry, in its irreducible formalism, may be merely the distant consequence of questions posed by powerless harpedonaptes (ancient Egyptian surveyors). And Canguilhem, retracing the history of the life sciences, adds that there is not a single fundamental knowledge in biology or physiology that cannot be traced back to its origin in a body in pain.[12]

Three Lessons

Actors involved in hybrid forums have been confronted with the question, which is difficult because it is new, of the representation of emergent minorities. By endeavoring to find answers, they have invented, developed, and put to the test equitable and efficient procedures, demonstrating in a concrete way the importance, indeed the preeminence, of procedural justice.

Procedural innovations intended to give existence to technical democracy have a domain of validity that undoubtedly extends beyond sociotechnical controversies. Procedures that prove to be equitable and efficient when it is a matter of opening political debate to groups of patients, residents of a site designated for storing nuclear waste, or farmers dis-turbed by the extension of transgenic cultivation may again prove to be so when what is at stake concerns ethnic or religious minorities. In this per-spective, three lessons could be pondered that suggest three lines of further exploration.

• The consideration of minority identities is more effective when their spokespersons are associated with the debates early on and in a continuous and productive way. Too often, the question of minorities is raised, and thus taken into consideration, only when the latter have formed estab-lished action groups. It is then difficult to negotiate their identity; opposi-tions have hardened, and confrontations have created strong resentments. The search for a common world becomes problematic. Why not take inspi-ration from one of the essential lessons of hybrid forums? This lesson is that identities are malleable when they emerge and, as a consequence of this, political debate is conceivable. We have seen that this presupposes great vigilance and attention to the weak signals that enable us to detect the emergence of identities lacking recognition.

• The rehabilitation of non-specialist competencies, and more precisely the competencies of concerned groups, in comparison with those of experts, could no doubt be tested on other issues than those with a marked scien-tific and technical component. It is probable, but it remains to be con-firmed, that debates on the composition of the collective are enriched and facilitated by such skills. Counterbalancing the power of all kinds of experts, not by the power of counter-experts, nor even by the organization of a pluralist expertise, but more radically by the early consultation of the "interested" through procedures inspired by those devised by hybrid fo-rums, could be one of the orientations to be favored in the reconfiguration of delegative democracy so as to make it more able to deal with the ques-tion of minorities. This could lead to a questioning of systematic recourse to "wise men" committees, which are often assisted by social science spe-cialists, and which multiply the sources of expertise, but without crossing the border and going so far as to set up a genuine consultation of the sup-posedly concerned groups. As Dewey notes, explorations and inquiries are the foundations of political processes. We have spoken a lot about the life sciences and nature, but secluded research in the social and human sciences should also open onto research in the wild on certain subjects cov-ered by these disciplines. The case of feminism and gender studies is a clear illustration of what such research collectives could be, and their role in the construction and recognition of the identities of emergent concerned groups.

• Another lesson, also linked to the question of the management of the tension between the general interest and particular interests, concerns the need to relativize general principles and standards in order to deal with questions that are always local, singular, and non-equivalent. Questions raised by farmers in the east of France cannot be answered by employing criteria identical to those used to deal with the issue of the nuclear waste site situated in the south, even if in both cases the farmers live in the vicinity of nuclear waste sites. One of the major lessons of hybrid forums is that the procedures must guarantee that the specificity of the questions, anxieties, and competences of the different concerned groups has been able to be expressed and taken into consideration. This result can be transposed unscathed to questions affecting religious or ethnic identity, or to ethical questions. Should wearing the chador be accepted? There are no grounds for deciding this question in general. The main thing in cases like this would be the definition of a procedure of consultation that leaves the widest place to local groups, facilitates their emergence and the expression of their points of view, and examines what is considered to be acceptable behavior in the limited framework of a particular college, school, or factory.

Hybrid forums multiply, and this multiplication underlines the limits of delegative democracy. Its institutions find it increasingly difficult to resist the overflows caused by science and technology. Groups are emerging that challenge its legitimacy, denouncing the monopoly of specialists and experts and also demanding a fairer representation of their identity. To achieve their goals, they are developing original procedures, putting them to the test, and endeavoring to draw lessons from them that will gradually enable technical democracy to exist.

One way of not hearing these voices and of ignoring the procedural innovations that introduce science and technology into democracy is to pretend to believe that these overflows and the demands to which they give rise can be solved by the existing institutions, and thus to reduce socio-technical controversies to simple questions of the management and negotiation of risks, of the adaptation of markets, or of the organization of expertise and its relations with decision makers.

If we wish to listen to the lessons given by all those who, by inventing technical democracy, reinvent democracy, we must abandon these conservative and defensive reflexes. What is at stake is obviously not the questioning of delegative democracy, but its enrichment. From this point of view, we should not underestimate the exemplary character of hybrid forums. The techno-sciences are a constant source of the renewal of identities.

They cause the social to "proliferate" and therefore pose the question of the representation of minorities in an acute way. All the procedures devised tend, with greater or lesser success, to bring satisfactory answers to this question.

The analysis of the innovations made in hybrid forums has enabled us to show the importance of the degree of dialogism of procedures. It has also led us to recognize that dialogic procedures are equitable procedures that lead to equitable—and efficient—measures. By demonstrating this series of equivalences, which found procedural justice, we have justified the possible transposition of these procedures. Thus goes the democratization of democracy, urged on by social actors and picked up by the social sciences in an endeavor to give some generality to the procedural innovations these actors have proposed.

Epilogue

Ladies and gentlemen, please, be serious at least for a few moments! Can you deny that it is thanks to science that our societies have made such formidable progress over the last decades? Are you so blind that you cannot see all the suffering secluded research has spared us, all the freedom it has given us, and all the well-being that we owe it? Have you forgotten that life expectancy, which was 25 years in 1789, is now 78, and this without an improvement in the human race? Do you really want us to quit the scene and give way to those prophets of gloom who wallow with delight in the imaginary catastrophes they predict for us, those ayatollahs who make pompous prophecies, establish a reign of unbearable intellectual terror, and, aided by journalists in search of the sensational, unleash a veritable collective hysteria? Do we want, do you want ladies and gentlemen, to return to the bronze age? If we were to listen to you and those like you, soon we would no longer be able to eat meat! We would be forced to use wood for heating, candles for lighting, or ox carts for transport! You are sated with progress. You are affluent! Poor spoilt children who enjoy breaking the toys you are given!

Pull yourselves together! Scientists must remain scientists, experts must remain experts, and politicians must not shirk their responsibility to make firm decisions on incommensurable options. I know that the fashion is for hybridization, for interbreeding. United Colors of Benetton! But allow me to resist this fashion! Allow me to be convinced that we need clearly drawn boundaries and that the ambiguities you have a liking for are the worst danger for our civilization! The devil, must I remind you of etymology, is the one who cuts across, divides, and creates confusion, he is an oblique being who wallows in ambiguity. Ah, ladies and gentlemen intellectuals, how the devil would love your hybrid forums that will very quickly be transformed into perfect tribunes for charlatans and quack doctors! I hear him coo with satisfaction when you speak of technical democracy. He is delighted with this monstrous coupling that mixes that which should absolutely be distinguished. He is pleased with your naive tolerance. He sees clearly that it is reason you are endangering and that soon there will only be real forgers and fake specialists. So stop playing with the devil! Return to the true values and to the forms of organization that have proved their effectiveness and that we have established with such difficulty. Do not destroy in one day what was built over centuries! The Republic needs true scientists, true experts, and true decision makers. To have done with uncertainties, which are certainly legitimate—who is not anxious in a world that is changing so rapidly?—we must redouble our efforts to inform, explain, and communicate. Join us in this task instead of denigrating and denying everything. We need your intelligence. But a sound, positive, and not diabolical intelligence!

Applause in the audience. It is a day like any other, a round table like any other, in a town like any other. An ordinary debate on an ordinary theme: the social impact of new technologies. Our interlocutor is just a bit more virulent than the others, and he has a bit more talent. But, visibly, he expresses what everyone here thinks and would like to have said. Delegative democracy still has some fervent supporters who are not without quality, strength, and talent.

Why such lack of understanding? Why such indignation? Why such harsh words, which seem to want only to encourage a bellicose atmosphere?

We could not find the words that would have enabled us to get out of this impasse and renew the dialogue. We did not have enough time. It is always difficult to avoid Manichaeism. That is why we decided to write this book. Perhaps in the calm of writing we will have more success in doing what we were unable to do in the heat of debating before an audience whose hostility we felt. This is our most cherished hope.

Will the reader have measured how many of the accusations proffered by our detractor were unjust? We venture to think so. This work is not a plea for a return to barbarism. Its main argument is that delegative democracy, which our interlocutor defends with talent and brio, is no longer enough to manage the innumerable overflows generated by the sciences and technology. Nowhere is it claimed that delegative democracy should be thrown out the window. The book proposes the establishment of a dialogic democracy that does not replace delegative democracy but enriches and nourishes it. Delegative democracy prospers and demonstrates its effectiveness when knowledge and identities are stabilized, but it must be supplemented when uncertainties and the controversies they feed take hold. Managing the tension between the hot and the cold by allowing uncertainties the space they need so that they can be transformed step by step into robust realities, and never interrupting this movement, is the sole principle that has motivated us. Neither a pure delegative democracy nor a pure dialogic democracy, but the combination of the two.

When we are confronted with uncertainties, two attitudes are in fact possible. The first attitude (which is somewhat pusillanimous) is to consider them as threats to be eliminated and reduced. The second (which is posi-

tive) is to recognize that they are a starting point for an exploration intended to transform and enrich the world in which we decide to live. The
first attitude is adopted spontaneously by all who are convinced that only
delegative democracy, notably because of its previous services, can manage
the social acceptability of science and technology. The second attitude,
which we have adopted, sees in the exploration and discussion of social,
scientific, and technical uncertainties the best means for arriving at an
always provisional, acceptable, and accepted order.

The thesis we are defending is that not only must existing controversies
be welcomed and recognized as participating in the democratization of democracy, but in addition they should be encouraged, stimulated, and
organized. There are overflows everywhere. They produce the fabric of our
individual and collective lives. They are everywhere, but generally they are
invisible. They spread insidiously, and when they become perceptible it is
often too late. That is why we have insisted on the importance of procedures that foster vigilance at every moment and are intended to identify
and explore overflows as soon as possible. We should not be content to
wait for controversies to break out. We should help them to emerge and
to become structured and organized. Controversies should be the constant
object of our concern. The constant preoccupations of technical democracy
are to facilitate the identification of concerned groups by themselves and
their partners and to organize collaborative research and the co-production
of knowledge that it makes possible. Dialogic democracy is not a conces-
sion, a stopgap. It nourishes representative democracy, and, once uncertainties have been reduced and the risks identified, it enables delegative
democracy to express all its effectiveness.

"We accept," our pugnacious interlocutor will concede, "that it is necessary to make concessions, organize debates, and permit people to become
aware and to express demands. But why exaggerate? Why wear oneself out
tracking down overflows even before they are visible? Cultivating disagreements and manufacturing uncertainties will end up dangerously rocking
the boat. Too much dialogic democracy will end up killing delegative democracy. Hybrid forums, agreed, but in homeopathic doses, and as a last
resort. The best strategy for containing repetitive crises is not to foster crisis
situations but to rarefy them. If all our energy is directed toward the exploration of overflows we will end up seeing them everywhere and seeing only
them!"

Our interlocutor would no doubt be right if these overflows, in their diversity and multiplicity, were to contribute to making up what he calls a
crisis situation. However, nothing is less certain.

Crisis of science? This is what some sociologists or philosophers assert. It is repeated over and over by a handful of scientists or experts who cry wolf because they love to present themselves as victims.

However, all the opinion surveys and inquiries prove that the image of science and of scientists has never been so well supported. A recent article in the newspaper *Le Monde*, titled "The French have faith in scientists," presents the results of an inquiry carried out on behalf of France's Minister of Research.[1] Like all previous surveys carried out on this subject, in France or in other countries, this one demonstrates that scientific research is a tremendous success with the good people (90 percent think that it should be a priority, and 30 percent even think it should be the government's main priority), and that the French have the highest opinion of scientists: "To control scientific progress and ensure respect for ethical questions, 53 percent of those questioned had faith first of all in the scientists themselves, far ahead of intellectuals and philosophers (19 percent) and religious authorities (6 percent)," write J.-F. Augereau and J.-P. Le Hir. It is entertaining to observe that those who ramble on *ad nauseam* about the public's loss of confidence in scientists are the ones the same public mistrust the most! Those surveyed think, moreover, that researchers practice an attractive profession, open to the world and socially valued. To say that there is a crisis of science or of research is therefore an overstatement. This public, whose opinion is wildly invoked, wants even more research and even more researchers to confront the great issues of health (84 percent), the environment (54 percent), and the supply of energy (32 percent). If there is a crisis, it is not a crisis of science or a crisis of research; it is a crisis of a shortage of research.

Since the crisis of science is a fable, should we look for a crisis of democracy? We might be tempted to think so. Does not the same survey say that the public mistrusts political men like the plague, and more generally political parties and all those well-established appointed spokespersons, intellectuals spreading in the media, rabbis, pastors, or other prelates *in partibus* who give moral lessons to whoever wants to listen? It is tempting to accept this diagnosis, since it is pleasant to let oneself be beguiled by the noxious fragrance exhaled by a discourse of crisis that constantly harasses us. In a regime that likes to think of itself as democratic, and calls itself democratic, what is more normal, more healthy, or less pathological than that the represented mistrust their representatives? Have we not repeated on every page that representation is a never-completed process in which there are, simultaneously, the person who speaks and the person in the name of whom he speaks and who is by this very fact reduced (at least for a while) to silence? There are not individuals already there, endowed with a will and with perfectly determinate preferences, which they transmit to their representatives, whose main quality is being faithful to their mandate. Representation is not a matter of loyalty. It does not aim to give a satisfactory image. It can only be provisional, only achieving felicity in the rare moments when consultation takes place and the spokesperson says to the person for whom he speaks "I say what you say." As soon as the break takes place— that is, as soon as the consultation comes to an end—betrayal is close. The representative must then present himself again to his constituents. This is not so much a matter of a lack of trust as a matter of a demand for the permanent reactivation of consultation. Democracy really would be in crisis if the public had blind confidence in its appointed representatives. By declaring that they are not prepared to delegate to its appointed representatives, without control, the task of orienting scientific progress and settling ethical questions, the represented, whether individuals or emergent groups, recall that representation exists only when there is consultation. They do not challenge the mechanisms of representation; they do not demand the establishment of a state in which, finally, representatives would be faithful. They know better than any philosopher that what matters are procedures of consultation that produce chatterboxes and without which the voices of some would have no other function than that of ensuring the silence of others. When, in the same inquiry, those who were surveyed expressly demand to be better and more frequently consulted on debated subjects such as genetic research, the modes of production of the food-processing industry, GMO, or the choice of energy policy, they are merely demanding more democracy, not acting against the limits of democracy. They are demanding a deepening and broadening of consultation without thereby denouncing the crisis of representation.

Wider, more diversified, more frequent, and deeper consultations—this takes us away from the stereotypical discourses on the need or the risks of a more direct or participatory democracy. The procedures of dialogic democracy do not introduce the people into the arena; they do not mean that a democracy captured by elites will suddenly turn into a more authentic regime. Since representation is not inscribed in the register of fidelity, there is no representative democracy that can do without the break between representatives and represented. To speak of direct democracy or of a return to the grassroots has no meaning. The sole argument of this book is that there are different regimes and modalities of consultation. The only question is that of the procedures put to work to organize this consultation. The only perspective is that of a dialogic democracy whose aim is to facilitate the exploration of possible worlds and the composition of the collective in situations where the procedures of delegative democracy manifest their limits. The development of dialogic democracy does not signify the end of representation, but its dissemination and proliferation. If we had to characterize it, we would say that it allows for the continuous expression of changing and emergent identities.

Our detractor might say "Let's accept for a moment that you are right, that the crisis of science or democracy is a pure invention, a historical nonsense. Let's imagine that we set up the procedures that you recommend, that we facilitate an intensification of collaborative research and the involvement of concerned groups, and that we extend the procedures of consultation. I still fail to understand why we need to turn our institutions upside down from top to bottom. Why would it not suffice to organize delegative democracy in terms of the needs and problems that arise? Nothing obliges us, as you suggest, to turn everything upside down, to accord as much importance to dialogic democracy as delegative democracy. Let's keep our heads and not let ourselves be too much influenced by some prions or transgenic plants. Let's be content with providing exceptional procedures for exceptional situations. Let's not run the risk of fostering the emergence and multiplication of these situations by devising institutions that need these situations to exist in order to function! You don't change a winning team, especially when it is a matter of dealing with scattered demands and possibly ephemeral unrest. It is better to wait. We live in a time of transition. Let's let things sort themselves out."

Our detractor is on the ball. Here is someone who does not hesitate to change tack. He sings us the grand song of crisis. He now speaks of simple troubles of growth, of episodes that are not grave. Fine rhetoric, especially since the argument is not without force. Antithetically, the euphemizing theme is more convincing to the same extent as the crisis was difficult to sustain. Faced with the daily avalanche of qualifications to describe the era into which we are supposed to be entering (post-modernity, the information society, the new economy), who can still believe in the alleged revolutions, historical changes, and other dramatizations dreamt up by ideologues craving notoriety? Why should we accept that there is a before and an after the Mad Cow crisis, or that the relations between science and politics should be thoroughly reexamined? Let's not lose our heads over events that may later prove to be mere epiphenomena!

We are forced to acknowledge that we do not have any formal proof of the increasing importance of socio-technical controversies. The wager of this book, for it is a wager, is the view that this movement is irrepressible.

The only sign (very weak, it is true) is that provided by the steady work that some people carry out day by day devising new consultative procedures and putting them to work. Intellectuals may be deceived; it is rarer for actors, without official support, and sometimes even against established powers, stubbornly to persist in error on a long-term basis. The ineluctable consequence of the growing success and increasingly central place occupied by science and technology is the proliferation of overflows and the increasing importance of the political problems they pose. Everywhere, on every front, concerned minorities appear who demand to be heard and who insist on being involved in the work of investigation. For science and research to continue to make their contribution to this collective exploration, there is no other solution than that of voluntarily and constantly promoting the establishment of technical democracy.

"Let's accept, at least in principle," our detractor retorts, "that this movement is ineluctable and that overflows are becoming omnipresent. Let's agree that dialogic democracy is assuming as much importance as delegative democracy. There is nonetheless something in your argument that does not add up. What do you mean when you claim that this dialogic democracy, which is all that you talk about, will enrich representative democracy? I tend to think, on the contrary, that delegative democracy and its assets are in danger of disappearing without a trace. Too much dialogic democracy, my dear ladies and gentlemen, will end up killing delegative democracy and, in the end, democracy *tout court*, just as bad money chases out good. In a nutshell, what I fear most is the tyranny of your minority groups. Let's listen to them when they emerge, agreed. But by taking them too much into consideration you will end up destroying individual rights, which, like so many other intellectuals of evil memory, you take lightly. Your dialogic democracy will quickly resemble a pitched battle between groups which think only of their own interests, struggling for power, and intolerant of any objections on the part of their members. You are prisoners of an insurmountable contradiction! How do you reckon to reconcile a collective made up of groups (dialogic democracy) and a collective made up of individuals endowed with inalienable rights (delegative democracy)? You have to choose. Or else the groups will choose for you, and with them tyranny will be installed."

The risk exists. But it is more imaginary than real, in the first place because delegative democracy is resistant to socio-technical controversies. Our political and technical culture and the institutions in which it is embodied multiply obstacles to the organization of hybrid forums. When the latter are set up, it is always afterwards, when the problems have become so difficult to deal with and manage that opening up debate remains the only conceivable outcome. If there is tyranny, it is not (yet) the tyranny of minorities! So let's begin by encouraging preventative sociotechnical controversies, those that are conceived of as tools for the exploration of uncertainties, before making us fearful with hypothetical abuses! All the more so since, apart from technological issues, there are many other issues, such as those of pension plans, internal security, or the cost of health care, to which the procedures devised for hybrid forums could be applied.

Let's be more precise. In a representative democracy that would accord dialogic democracy the place it deserves, the risk of the tyranny of minorities would be lessened as soon as dialogic procedures were devised to make spokespersons revocable whenever they tended to silence in the long term those in whose name they speak. In order that identities can be negotiated and consolidated, consultation within the minority groups should be renewed frequently. But this is not the main thing. If hybrid forums are an enrichment of delegative democracy, and not a threat to it, it is because in practice they make it possible to get out of a theoretical contradiction that political philosophy comes up against. They replace a conception of the public space made up of detached, transparent actors lacking existential substance with a "cluttered" public space in which individual wills are worked out and nourished by attachments that concerned groups have negotiated and discussed at length and in breadth.

The public space of hybrid forums is not reducible to that imagined by Hannah Arendt, Jürgen Habermas, or John Rawls. Like the public spaces of Habermas and Arendt, it privileges debate, discussion, the exchange of arguments, and the will of everyone to understand and listen to each other. Like that of Rawls, it puts on the stage actors who are not solely concerned with their own interest, but who are involved, through the very logic of the procedures followed, in relations of reciprocity: each takes the others' points of view into account. But, unlike the public spaces described by these three authors, it does not specify that the participants be persons or individuals divested of every particular quality and detached from their networks of sociability, having bracketed off everything they value and everything to which they are attached, that is to say, everything that makes up their irreducible identity, including their bodies, genes, and emotions, which sometimes prevent them from speaking and taking part in the public debate! Of course, neither Arendt nor Habermas nor Rawls suggests that the consideration of the positions and interests of each must be discussed by individuals purged of all substance. They acknowledge the multiplicity of identities and the irreducible differences that separate persons. This is, moreover, why they consider the calculation of interests to be insufficient and see the need to devise a set of arrangements that enables everyone to take all the others into account. But the paradox is that, for all these authors, this movement is possible only if individuals have been freed from their attachments before entering the debate. We are familiar with the strange definition proposed by Rawls:

Somehow we must nullify the effects of specific contingencies which put men at odds and tempt them to exploit social and natural circumstances to their own advantage. Now in order to do this I assume that the parties are situated behind a veil of ignorance. They do not know how the various alternatives will affect their own particular case and they are obliged to evaluate principles solely on the basis of general considerations. It is assumed, then, that the parties do not know certain kinds of particular facts. First, no one knows his place in society, his class position or social status; nor does he know his fortune in the distribution of natural assets and abilities, his intelligence and strength, and the like. Nor, again, does anyone know his conception of the good, the particulars of his rational plan of life, or even the special features of his psychology such as his aversion to risk or liability to optimism or pessimism.[2]

In the same way, Arendt requires the persons who discuss the common good to be freed from all material contingencies.[3] And Habermas imagines human beings entirely absorbed in their will to communicate. Since Kant, political philosophy tends to consider it necessary that, before entering the public space, human subjects be severed from all the attachments that hold them in the world and be stripped of their own bodies, social identities, and existential problems so that they are no longer interested in anything but the common good. The concern for justice requires these transparent beings, who are rational by dint of being transparent, and who decide not to discuss the good, that which they value and to which they are attached, so as to be able to concentrate solely on questions of justice. The just comes before the good. These authors' recognition of the existence of singularities, differences, and attachments, and their attempt to devise procedures for removing them, make their position broadly unrealistic. Furthermore, they nourish the notably relativist critics who denounce the concealment of relations of force behind the image of persons debating questions of justice and equity in a disinterested and disembodied way.

The experiences described in chapters 4 and 5 show us that the need for such detachment and de-socialization exists only in philosophy manuals. Dialogic procedures, whose equitable characteristics we have noted, do not require disembodied beings. They contribute, on the contrary, to the constitution of a public space inhabited by cluttered, attached "selves" who can exchange views with and understand one another only because they are, precisely, attached and cluttered. The protagonists arrive with their damaged genes, with their anxieties as residents concerned about toxic discharges, or with their demands concerning the organization of drug trials, attached "selves" who can exchange views with and understand one another only because they are, precisely, attached and cluttered.

Those with myopathy, phase 1: To live happily, they live hidden away, relegated to the private sphere. Those most affected are monsters that the families dare not expose to the gaze of others. Those with myopathy, phase 2: After lengthy collective work, they expose themselves. And if they do so, it is thanks to their investment in research, both secluded and in the wild, which has enabled them to recompose and reconstruct their identity and to introduce it into the public space. This to-and-fro movement between the exploration of possible worlds and the exploration of identities with a view to their composition, demonstrates how, in practice, an individual or a group can lay claim to a cluttered, constituted identity (in this case, through either mutated or absent genes) and can also take part in a public debate in which these cluttered identities are the central object of the discussions. At no time are the specific characteristics of those with myopathy bracketed off, killed, or referred back to the private sphere. The hybrid forum, in its very organization, tends instead to put an end to this separation. It is as persons suffering from myopathy that the persons struck by neuromuscular diseases, unable to breath and sometimes even to speak, enter the public space. Will we ask them to leave at home their wheelchairs and the cannulas that inflate their lungs, out of fear that their judgment may be biased? Will we require them to keep quiet about their genes and the drugs they would like to see developed, out of fear that their egoism will interfere with the pursuit of the common good? Not only would this condemn them to non-existence; it would prevent a society being debated, evoked, or desired in which all those who have damaged bodies live with the same rights and duties as their fellow men and women. It would be to refrain from broadening and enriching conceptions of the human person. Each, by discussing what is good for him or her, discusses what is good for the other. Justice and the good combine together in the same movement. How can we take account of emergent identities, so as to adjust them to one another by transforming them, if these identities are not discussed?

From these hybrid forums come richer beings and more complex and open communities. When the ordinary citizens of delegative democracy come to elect their representatives, they will be able to choose from within a wider range of collectives, including all those that propose the same possibilities of existence for each, whatever the disabling situations in which they find themselves. If these hybrid forums had not taken place, if dialogic procedures had not been followed, it would not have been possible to have a conception of the human person that also includes the limits imposed by a bad adjustment between the body and the possibilities it is offered, or alternatively the option of choosing the risks that one incurs or makes others incur. Yes, dialogic democracy enriches representative democracy. And if it does so, it is because the uncertainty it confronts is considerable and because nothing is fixed *a priori*. None of the contingencies and attachments that form the substance of the human being, and over which Rawls demands that we throw a veil, are stabilized, fixed, and attached to any beings whomsoever. Thanks to collaborative research, identities are explored at the same time as the collective. In these conditions, why hide and withdraw from debate precisely that which we should be able to discuss so that it might be modified and redistributed?

In hybrid forums, attachments and entanglements make communication possible, rather than prevent it. This is true to the extent that getting debate underway does not require individuals endowed with speech and able to engage in oratorical jousts. Relations of reciprocity are established between parents and the children they care for, who often are unable to articulate well-formed sentences. Such genuine communication does not need words and embodies a common humanity reinvented by each in order later to be debated in the public space. The public space of hybrid forums, a public space cluttered by beings who are themselves cluttered and attached, constitutes an irreplaceable laboratory in which our common humanity and the communities compatible with it are redefined at the same time. "No, sir, the devil is not in the hybrid forums," because, as Michael Sandel writes, "To imagine a person incapable of constitutive attachments ... is ... to imagine a person wholly without character, without moral depth."[4] This also means, we would add, a refusal to see that what constitutes our common humanity must be permanently tested and collectively debated.

Notes

Chapter 1

1. National Radioactive Waste Management Agency (Agence nationale pour la gestion des déchets radioactifs).

2. On the genesis and implementation of this law, see Yannick Barthe, *Le Pouvoir d'indécision. La mise en politique des déchets nucléaires* (Economica, 2006).

3. Élisabeth Rémy, "Comment dépasser l'alternative risque réel, risque perçu?" *Annales des Mines*, series "Responsabilité et environnement," no. 5 (January 1997): 27–34.

4. See Michel Callon and Arie Rip, "Humains, non-humains: morale d'une coexistence," in *La Terre outragée*, ed. J. Theys and B. Kalaora (Autrement, 1992).

5. Marilyn Strathern, "What is intellectual property after?" in *Actor Network Theory and After*, ed. J. Law and J. Hassard (Blackwell, 1999).

6. Many works have demonstrated this. See Arie Rip, "Controversies as Informal Technology Assessment," *Knowledge: Creation, Diffusion, Utilization* 8, no. 2 (1986): 349–371; Alberto Cambrosio and Camille Limoges, "Controversies as Governing Processes in Technology Assessment," *Technology Analysis and Strategic Management* 3, no. 4 (1991): 377–396.

7. Anne-Marie Querrien, "Un tournant dans la pratique de la concertation: le TGV Méditerranée," *Annales des Ponts et Chaussées* no. 81, 1997: 16–23.

Chapter 2

1. Christian Licoppe, *La Formation de la pratique scientifique. Le Discours de l'expérience en France et en Angleterre, 1630–1820* (La Découverte, 1996).

2. We use this word to designate the *experimentum*, the construction that produces unusual phenomena.

3. On the notion of translation, see Michel Callon, "Some Elements of a Sociology of Translation: Domestication of the Scallops and the Fishermen of St Brieuc Bay," in *Power, Action and Belief*, ed. J. Law (Routledge & Kegan Paul, 1986); Bruno Latour, *Science in Action: How to Follow Scientists and Engineers through Society* (Harvard University Press, 1987); Michel Callon, ed., *La Science et ses réseaux. Genèse et circulation des faits scientifiques* (La Découverte, 1989).

4. On the AFM (Association française contre les myopathies) see Vololona Rabeharisoa and Michel Callon, *Le Pouvoir des malades. L'Association française contre les myopathies et la recherche* (Presses de l'École des mines de Paris, 1999); Michel Callon and Vololona Rabeharisoa "The Growing Engagement of Emergent Concerned Groups in Political and Economic Life. Lessons from the French Association of Neuromuscular Disease Patients," *Science, Technology and Human Values* 33 (2008), no. 2: 230–261.

5. On the notion of inscription, and more generally on the description of this inversion of the relation of forces, see Bruno Latour and Steve Woolgar, *Laboratory Life: The Construction of Scientific Facts* (Princeton University Press, 1986); Latour, *Science in Action*.

6. Alberto Cambrosio and Peter Keating, *Exquisite Specificity. The Monoclonal Antibody Revolution* (Oxford University Press, 1995).

7. Gaston Bachelard, *Le Matérialisme rationnel* (Presses universitaires de France, 1987 [1953]), p. 10.

8. For an overview see Michel Callon, "Four Models for the Dynamic of Science," in *Handbook of Science and Technology Studies*, ed. S. Jasanoff et al. (Sage, 1995).

9. Gaston Bachelard, *L'Activité rationaliste de la physique contemporaine* (Presses universitaires de France, 1951), p. 113; Bruno Latour, "When Things Strike Back. A Possible Contribution of 'Science Studies' to the Social Sciences," *British Journal of Sociology* 51, no. 1 (2000): 107–123.

10. Edwin Hutchins, *Cognition in the Wild* (MIT Press, 1992).

11. Sophie Houdart, *La cour des miracles. Ethnologie d'un laboratoire japonais* (Editions du CNRS, 2008).

12. Robert Köhler, *Lords of the Fly: Drosophila Genetics and the Experimental Life* (University of Chicago Press, 1994).

13. National Center of Scientific Research (*Centre national de la recherche scientifique*).

14. Atomic Energy Commission (Commissariat à l'Energie Atomique).

15. National Institute of Agronomic Research (*Institut national de la recherche agronomique*).

16. National Institute of Health and Medical Research (Institut national de la santé et de la recherche médicale).

17. See Michel Callon, "The State and Technical Innovation: A Case Study of the Electrical Vehicle in France," *Research Policy* 9, no. 4 (1980): 358–376.

18. Délégation générale à la recherche scientifique et technique (the state body responsible for scientific and technical research).

19. Madeleine Akrich, "The De-Scription of Technical Objects," in *Shaping Technology/Building Society*, ed. W. Bijker and J. Law (MIT Press, 1992).

20. Bruno Latour, *The Pasteurization of France* (Harvard University Press, 1988); Bruno Latour, *Le Métier de chercheur. Regard d'un anthropologue* (INRA Éditions, 1995).

21. Represented by the dotted arrow in figure 2.3.

Chapter 3

1. Steven Epstein, *Impure Science: AIDS, Activism, and the Politics of Knowledge* (University of California Press, 1996).

2. Robin Horton, "African Traditional Thought and Western Science," in *Rationality*, ed. B. Wilson (Blackwell, 1970).

3. Phil Brown, "Popular Epidemiology and Toxic Waste Contamination: Lay and Professional Ways of Knowing," *Journal of Health and Social Behavior* 33, no. 3 (1992): 267–281; Phil Brown, Sabrina McCormick, Brian Mayer, Stephen Zavestoski, Rachel Morello-Frosch, Rebecca Gasior Altman, and Laura Senier, "'A Lab of Our Own': Environmental Causation of Breast Cancer and Challenges to the Dominant Epidemiological Paradigm," *Science, Technology, and Human Values* 31, no. 5 (2006): 499–536.

4. E. E. Evans-Pritchard, *Witchcraft, Oracles, and Magic among the Azande* (Clarendon, 1976).

5. Jeanne Favret-Saada, *Deadly Words: Witchcraft in the Bocage* (Cambridge University Press, 1981).

6. Euripides, *Medea*, in Euripides, *Cyclops. Alcestis. Medea*, ed. D. Kovacs (Harvard University Press, 1994). [The English translation has been slightly modified to bring it into line with the French translation cited by the authors: *Médée*, ed. L. Méridier, volume 1, fourth revised and corrected edition (Les Belles Lettres, 1956), p. 123—G.B.]

7. Rabeharisoa and Callon, *Le Pouvoir des malades*; Rabeharisoa and Callon, "The Involvement of Patients in Research Activities Supported by the French Muscular Dystrophy Association," in *States of Knowledge*, ed. S. Jasanoff (Routledge, 2004).

8. All the techniques linked to procreation and its control.

9. Adele E. Clarke, *Disciplining Reproduction. Modernity, American Life Sciences, and The Problems of Sex* (University of California Press, 1998).

10. See Steven Epstein, "The Construction of Lay Expertise: AIDS Activism and the Forging of Credibility in the Reform of Clinical Trials," *Science, Technology, and Human Values* 20, no. 4 (1995), p. 416.

11. Michael Pollak, *Les Homosexuels et le sida. Sociologie d'une épidémie* (A.-M. Métailié, 1988).

12. Janine Barbot, "Agir sur les essais thérapeutiques. L'expérience des associations de lutte contre le sida in France," *Revue d'épidémiologie et de santé publique* 46 (1998): 305–315.

13. Brian Wynne, "May the Sheep Safely Graze? A Reflexive View of the Expert-Lay Knowledge Divide," in *Risk, Environment and Modernity*, ed. S. Lash et al. (Sage, 1996).

14. Plato, *The Republic*, book V.

15. This episode is described in detail by Bernadette Bensaude-Vincent, *L'Opinion publique et la science. À chacun son ignorance* (Les Empêcheurs de penser en rond, 2000), p. 52, and in Léon Chertok and Isabelle Stengers, *Le Cœur et la Raison. L'hypnose en question de Lacan à Lavoisier* (Payot, 1989). In her book, Bernadette Bensaude-Vincent clearly shows how secluded research was progressively constructed by distancing itself from a public defined as ignorant and having to be educated.

Chapter 4

1. Gabrielle Hecht, *The Radiance of France. Nuclear Power and National Identity after World War II* (MIT Press, 1998), p. 56.

2. Jean-Jacques Rousseau, *The Social Contract*, in *The Social Contract and Discourses* (J. M. Dent, 1973), p. 185.

3. The most accomplished form of this gap is General de Gaulle's famous statement to the French in Algeria who were fighting against that country's independence: "I have understood you." The most important thing was said in four words. De Gaulle constituted himself as legitimate spokesperson (able to express the will of each citizen), gave existence to the will of the people by expressing it (he does not say, "here is what I am going to say and what you say," but leaves the content of this will undetermined, merely asserting that he is in possession of it, his speech being authorized by that very fact), and reduced it to silence ("since I have understood you, you no longer have to say anything: I am your instrument of phonation"). Only a few years later was it possible to understand ... what he had understood.

4. See chapter 7 for a more precise definition of representative democracy.

5. Alain Boureau, "L'adage *Vox populi, vox Dei* et l'invention de la nation anglaise (VIIIe-XIIe siècle)," *Annales. Histoire, Sciences Sociales* 47, no. 4 (July-October 1992): 1071–1089.

6. The concept of delegative democracy has another meaning as well, when it is used to characterize certain South American regimes where the Presidents of the Republic, elected by universal suffrage, are left free to govern as they deem fit once they are in power.

7. This separation is converted into a multitude of derivative divisions, like that between consumers and producers, etc.

8. The word 'composition' designates the action of composing as much as its result.

9. In delegative democracy, political logic is similar to a hunt for votes; moreover, the modalities of counting and aggregating votes have a strategic importance (modalities of voting).

10. The aim of these procedures is that by successive reductions only one can speak legitimately in the name of all.

11. For Rousseau, diversity makes individuals reasonable by forcing them to seek out what they have in common and which, being freely laid down, is imposed on each.

12. Charles Taylor, *Multiculturalism: examining the politics of recognition*, edited and introduced by Amy Gutman (Princeton University Press, 1992).

13. Ibid.

14. Georges Charpak, "Pour raison nucléaire garder, vive le DAIRI," *Le Monde*, 2 June 2000.

15. The French phrase is "remettre les compteurs à zero," which literally means "reset the meters at zero," but commonly means "start from scratch again," "wipe the slate clean," "go back to square one," etc.—G.B.

Chapter 5

1. See, however, the special issue of the journal *Science and Public Policy* 26, no. 5 (1999) on public participation in scientific and technological matters, coordinated by Simon Joss, a specialist on consensus conferences. Also see two articles of synthesis: Daniel J. Fiorino, "Citizen Participation and Environmental Risk: A Survey of Institutional Mechanisms," *Science, Technology, and Human Values* 15, no. 2 (Spring 1990): 226–243; Gene Rowe and Lynn J. Frewer, "Public Participation Methods: A Framework for Evaluation," *Science, Technology, and Human Values* 25, no. 1 (Winter 2000): 3–29.

2. See Aidan Davison, Ian Barns, and Renato Schibeci, "Problematic Publics: A Critical Review of Surveys of Public Attitudes to Biotechnology," *Science, Technology, and Human Values* 22, no. 3 (Summer 1997): 317–348.

3. See Rowe and Frewer, "Public Participation Methods: A Framework for Evaluation."

4. Dominique Memmi, "Celui qui monte à l'universel et celui qui n'y monte pas. Les voies étroites de la généralisation 'éthique,'" in *Espaces publics mosaïques*, ed. B. François and É. Neveu (Presse universitaires de Rennes, 1999). She notes that a nurse has greater difficulty detaching herself from her patients, whom she cares for and knows, than a doctor of the liver, who has more difficulties expressing himself in the name of the human person in general than a well-known professor who is accustomed to public occasions. It is certainly difficult to speak in the name of abstract beings, but it is less so than being the spokesperson of both these abstract entities and of particular patients: the work of legitimation of the ascent in generality must be constantly started again.

5. For an interesting analysis of focus groups see Javier Lezaun, "A Market of Opinions: the Political Epistemology of Focus Groups," in *Market Devices,* ed. M. Callon et al. (Blackwell, 2007).

6. This thesis is defended by those in charge of two experimental programs begun at the end of the 1990s. A European program (ULYSSE) and a Swiss program (CLEAR) are involved, both of them bearing on environmental questions and designed with the aim of developing deeper knowledge of the conditions of utilization of *focus groups*.

7. Jean-Michel Fourniau, "Information, Access to Decision-making and Public Debate in France: The Growing Demand for Deliberative Democracy," *Science and Public Policy* 28, no. 6 (2001): 441–451.

8. Michel Gariépy, "Toward a Dual-Influence System: Assessing the Effects of Public Participation in Environmental Impact Assessment for Hydro-Quebec Projects," *Environmental Impact Assessment Review* 11, no. 4 (1991): 353–374; Éric Montpetit, "Public Consultations in Policy Network Environments: The Case of Assisted Reproductive Technology Policy in Canada," *Canadian Public Policy—Analyse de Politiques* 29, no. 1 (2003): 95–110; Louis Simard, "'Preparing' and 'Repairing' Public Debate: Organizational Learning of Promoters in Environmental and Energy Governance," *Revue Gouvernance* 2, no. 2 (Fall 2005): 7–18.

9. Cécile Blatrix, "Le maire, le commissaire enquêteur et leur 'public'. La pratique politique de l'enquête publique," in *La démocratie locale*, ed. CURAPP/CRAPS (Presses universitaires de France, 1998).

10. This development is shared by a number of countries, above all by those with an Anglo-Saxon tradition. Examples include the Quebec BAPE and Australia. Projects are investigated by an inquiry commissioner who is a magistrate. He or she organizes public hearings in which the different opposing interests confront one another in a judicial mode. The commissioner may appeal to experts on the most controversial points, and he or she acts finally as a conciliator-judge in order to reconcile the opposing positions.

11. Simon Joss and John Durant, eds., *Public Participation in Science: The Role of Consensus Conferences in Europe* (Science Museum, 1995).

12. See Daniel Boy, Dominique Donnet-Kamel, and Philippe Roqueplo, "Un exemple de démocratie délibérative: la conférence française de citoyens sur l'usage des organismes génétiquement modifiés en agriculture et dans l'alimentation," *Revue française de science politique* 50, no. 4–5 (August-October 2000): 779–809.

13. Ibid.

14. Pierre-Benoit Joly et al., *L'Innovation controversée: le débat public sur les OGM en France* (INRA, 2000).

15. George Horming, "Citizens' Panels as a Form of Deliberative Technology Assessment," *Science and Public Policy* 26, no. 5 (1999): 351–359.

16. Les Levidow, "Democratizing Technology—or Technologizing Democracy? Regulating Agricultural Biotechnology in Europe", *Technology in Society* 20, no. 2 (1998): 211–226.

17. See Janine Barbot, "How to Build an 'Active' Patient? The Work of AIDS Associations in France," *Social Science and Medicine* 62, no. 3 (2006): 538–551; Janine Barbot and Nicolas Dodier, "Multiplicity in Scientific Medicine: The Experience of HIV-Positive Patients," *Science, Technology, and Human Values* 27, no. 3 (Summer 2002): 404–440. For a North American history of this "impure" science, see: Epstein, *Impure Science*.

18. For example, *zapping* (spectacular actions with strong media coverage), or the threat of *outing* (public revelation of the sexual orientation of public figures).

19. Respectively, the Conseil national du sida, the Agence française de lutte contre le sida, and the Agence nationale de recherche sur le sida.

20. The French Téléthon is very different to the US one. The event itself includes many festive activities in which a strong feeling of national solidarity is expressed (some have referred to it as the equivalent of Bastille Day celebrations, in winter), as well as numerous debates on biological research, its progress and its problems.

21. Had we wanted to provide an exhaustive inventory of available procedures, we would have had to present deliberative polling (see http://www.la.utexas.edu/research/delpol/) which provides a particularly ingenious compromise between the classical opinion polls, the German "Plannungzelle", consensus conferences, and citizen juries. As the reader will have noticed, our aim is not to be exhaustive. We have simply wanted to suggest the diversity of existing procedures and to show the interest of the evaluation criteria proposed in choosing procedures or designing them.

Chapter 6

1. *La Recherche*, no. 325 (November 1999), p. 5. (*La Recherche* is a French monthly magazine equivalent to *Scientific American*.)

2. G. Poste, *Financial Times* and *Courrier international* no. 480 (January 2000), p. 37.

3. Daniel Bodansky, "Scientific Uncertainty and the Precautionary Principle," *Environment* 33, no. 4–5 (1991): 43–44; Timothy O'Riordan and James Cameron, ed., *Interpreting the Precautionary Principle* (Earthscan, 1994); David Freestone, Ellen Hey, ed., *The Precautionary Principle and International Law. The Challenge of Implementation* (Kluwer Law International, 1996).

4. L. Sedel, "L'Assistance publique de Paris se meurt," *Le Monde*, 2 February 2000.

5. The first cases were identified in 1985, and it was in November 1986 that some vets alerted the Minister of Agriculture on the threat of an epidemic. Patrick van Zwanenberg and Erik Millstone, "'Mad Cow Disease' 1980s–2000: How Reassurances Undermined Precaution," in *Late Lessons from Early Warnings: The Precautionary Principle 1896–2000*, Environmental Issue Report 22, European Environmental Agency, 2001.

6. In Great Britain, the first measure of protection for exposed workers was in 1931 (1945 in France) and the first compensation for pathologies due to asbestos in 1933 (1950 in France).

7. Seven countries, including Germany, Holland, Italy, the Scandinavian countries, and Switzerland, adopted the measure of prohibition before France.

8. Michel Setbon, *Pouvoirs contre sida. De la transfusion sanguine au dépistage: décisions et pratiques en France, Grande-Bretagne et Suède* (Seuil, 1993).

9. Those who have spent time in the Caribbean, who state they have had homosexual relations, or who have taken drugs intravenously are excluded from giving blood.

10. An active approach adopted by the Californian law of 1986 called "proposition 65" for products suspected of having carcinogenic effects: the onus is on the company that wants to use a suspected molecule, to carry out research on the levels of exposure and the types of risk. See William S. Pease, "Identifying Chemical Hazards for Regulation," *Risk: Issues in Health and Safety*, no. 2 (1992): 127–172.

11. Eberhard Bohne and Günter Hartkopf write: "The environment policy is not limited to dealing with imminent threats and reducing damage sustained. Beyond, an environment policy of precaution requires that the basic natural elements are protected and attentively taken care of." (*Umweltpolitik, Grundlagen, Analysen und Perspektiven*, Westdeutscher Verlag, 1983, pp. 98–108) The notion of "basic natural elements" (*éléments naturels de base*) refers on the one hand to the sparing use of renewable natural resources and, on the other, to the least emission possible of polluting substances. See also Sonja Boehmer-Christiansen, "The Precautionary Principle in Germany," in *Interpreting the Precautionary Principle*.

12. Hans Jonas, *The Imperative of Responsibility: In Search of an Ethics for the Technological Age* (University of Chicago Press, 1985 [1979]).

13. Ibid.

14. Ibid.

15. Michael D. Rogers, "Risk Analysis under Uncertainty. The Precautionary Principle, and the New EU Chemical Strategy," *Regulatory, Toxicology and Pharmacology* 37, no. 3 (2003): 370–381.

16. Report of the session of CADAS 8 February 2000, p. 3.

17. Bernard Glorion, editorial, *Bulletin de l'Ordre des médecins* no. 10, December 1999.

18. Commissariat général du plan, working group on the perspectives for France, *Risques et Développement durable*, December 1999, p. 18.

19. Conseil d'État, *Réflexions sur le droit de la santé*, public report no. 49 (La Documentation française, 1998).

20. Didier Sicard, "Le danger, c'est de faire du principe de précaution une sorte d'imprécation," *Le Monde*, 22 December 1999.

21. Philippe Kourilsky and Geneviève Viney, eds., *Le principe de précaution*, Rapport au Premier ministre (Odile Jacob, 2000), p. 167. On all the issues mentioned here, the reader is referred to this very complete volume.

22. With the other principles: prevention of effects/damages by correction at the source, the principle that the polluter pays, and informing the public.

23. European Commission, DG XXIV, *Lignes directrices pour l'application du principe de précaution*, 17 October 1998.

24. European Commission, *Communication from the Commission of 2 February 2000 on the precautionary principle*, Brussels, 2 February 2000.

25. Olivier Godard, "L'ambivalence de la précaution et la transformation des rapports entre science et décision," in *Le Principe de précaution dans la conduite des affaires humaines*, ed. O. Godard (Éditions de la MSH/INRA, 1997).

26. Twenty-three cases were recorded, of which seven were mortal, at the end of February 2000.

27. Francis Chateauraynaud and Didier Torny, *Les sombres précurseurs. Sociologie pragmatique de l'alerte et du risque* (Editions de l'EHESS, 1999), pp. 76–78.

28. STSE: subacute transmissible spongiform encephalopathies.

29. Marie-Angèle Hermitte and Dominique Dormont, "Propositions pour le principe de précaution à la lumière de la vache folle," in *Le principe de précaution*, pp. 343–350.

30. European Commission, *Communication from the Commission of 2 February 2000 on the precautionary principle*, Brussels, 2 February 2000.

31. Clark Miller and Paul N. Edwards, ed., *Changing the Atmosphere: Expert Knowledge and Environmental Governance* (MIT Press, 2001); William F. Ruddiman, *Plows, Plagues and Petroleum: How Humans Took Control of Climate* (Princeton University Press, 2005)

32. Marie-Angèle Hermitte, "Le principe de précaution à la lumière du drame de la transfusion sanguine en France," in *Le Principe de précaution dans la conduite des affaires humaines*, p. 195.

33. Marie-Angèle Hermitte and Dominique Dormont, "Propositions pour le principe de précaution à la lumière de la vache folle," p. 355.

34. Ibid., p. 351.

35. This type of judgment (called "standard") should be understood a legal and open model of behavior which must be redefined in every case. It is a matter of a hypothetical model whose content must be constructed every time it is employed. The law regularly resorts to such fictions that constitute useful instruments of regulation and for resolving conflicts. Thus, "glaring error," "serious misconduct," "accepted standards of behavior," "normal rental value," and "serious dispute" are classical examples of standards whose content is today more or less stabilized by jurisprudence. These legal categories share three particular features. First of all, they contain a space of indetermination intentionally placed within the rule; this then calls for a specific activity of evaluation; and finally, this interpretation is based on facts both internal and external to the law. All recourse to a standard thus calls for a hybrid reasoning mixing legal and non-legal criteria.

Chapter 7

1. Ulrich Beck, *Risk Society: Towards a New Modernity* (Sage, 1992).

2. Philippe Roqueplo, *Entre savoir et décision, l'expertise scientifique* (INRA Éditions, 1997).

3. Quoted by Dominique Memmi, "Celui qui monte à l'universel et celui qui n'y monte pas. Les voies étroites de la généralisation 'ethique,'" p. 157.

4. Bruno Latour, *Politics of Nature: How to Bring the Sciences into Democracy* (Harvard University Press, 2004).

5. We cannot help thinking of Adam Smith's famous phrase in the first chapter of *The Wealth of Nations* (Oxford University Press, 1976, volume 1, book 1, chapter 1, p. 21) in which he announced this necessary seclusion for the spread of "the ingenuity ... of those who are called philosophers or men of speculation, whose trade it is, not to do any thing [*sic*], but to observe every thing; and who, upon that account, are often capable of combining together the powers of the most distant and dissimilar objects."

6. See Jean Gadrey, *New Economy, New Myth?* (Routledge, 2002).

7. Christian Delacampagne, *La Philosophie politique aujourd'hui. Idées, débats, enjeux* (Seuil, 2000), pp. 19–21.

8. John Dewey. *The Public and its Problem* (Ohio University Press, 1954); Walter Lippmann, *The Phantom Public* (Transaction, 1927).

9. R. Folger, "Reformulating the preconditions of resentment: A referent cognitions model," in *Social Comparison, Social Justice, and Relative Deprivation,* ed. J. Masters and W. Smith (Erlbaum, 1987).

10. For a general survey, see Simon Joss and Arthur Brownlea, "Considering the concept of procedural justice for public policy- and decision-making in science and technology," *Science and Public Policy* 26, no. 5 (1999): 321–330.

11. Barthe, *Le Pouvoir d'indécision.*

12. G. Canguilhem, *Le normal et le pathologique* (PUF, 1966).

Epilogue

1. J.-F. Augereau and J.-P. Le Hir, *Le Monde,* 30 November 2000, p. 26.

2. John Rawls, *A Theory of Justice* (Oxford University Press, 1971), pp. 136–137.

3. "The distinction between the private and the public realms ... equals the distinction between things that should be shown and things that should be hidden."— Hannah Arendt, *The Human Condition* (University of Chicago Press, 1958), p. 72.

4. Michael Sandel, *Liberalism and the Limits of Justice* (Cambridge University Press, 1998, second edition), p. 179.

Index

Inside Technology
edited by Wiebe E. Bijker, W. Bernard Carlson, and Trevor Pinch

Gabrielle Hecht, *The Radiance of France: Nuclear Power and National Identity after World War II*

Kathryn Henderson, *On Line and On Paper: Visual Representations, Visual Culture, and Computer Graphics in Design Engineering*

Christopher R. Henke, *Cultivating Science, Harvesting Power: Science and Industrial Agriculture in California*

Christine Hine, *Systematics as Cyberscience: Computers, Change, and Continuity in Science*

Anique Hommels, *Unbuilding Cities: Obduracy in Urban Sociotechnical Change*

Deborah G. Johnson and Jameson W. Wetmore, editors, *Technology and Society: Building our Sociotechnical Future*

David Kaiser, editor, *Pedagogy and the Practice of Science: Historical and Contemporary Perspectives*

Peter Keating and Alberto Cambrosio, *Biomedical Platforms: Reproducing the Normal and the Pathological in Late-Twentieth-Century Medicine*

Eda Kranakis, *Constructing a Bridge: An Exploration of Engineering Culture, Design, and Research in Nineteenth-Century France and America*

Christophe Lécuyer, *Making Silicon Valley: Innovation and the Growth of High Tech, 1930–1970*

Pamela E. Mack, *Viewing the Earth: The Social Construction of the Landsat Satellite System*

Donald MacKenzie, *An Engine, Not a Camera: How Financial Models Shape Markets*

Donald MacKenzie, *Inventing Accuracy: A Historical Sociology of Nuclear Missile Guidance*

Donald MacKenzie, *Knowing Machines: Essays on Technical Change*

Donald MacKenzie, *Mechanizing Proof: Computing, Risk, and Trust*

Maggie Mort, *Building the Trident Network: A Study of the Enrollment of People, Knowledge, and Machines*

Peter D. Norton, *Fighting Traffic: The Dawn of the Motor Age in the American City*

Helga Nowotny, *Insatiable Curiosity: Innovation in a Fragile Future*

Ruth Oldenziel and Karin Zachmann, editors, *Cold War Kitchen: Americanization, Technology, and European Users*

Nelly Oudshoorn and Trevor Pinch, editors, *How Users Matter: The Co-Construction of Users and Technology*

Shobita Parthasarathy, *Building Genetic Medicine: Breast Cancer, Technology, and the Comparative Politics of Health Care*

Trevor Pinch and Richard Swedberg, editors, *Living in a Material World: Economic Sociology Meets Science and Technology Studies*

Paul Rosen, *Framing Production: Technology, Culture, and Change in the British Bicycle Industry*

Susanne K. Schmidt and Raymund Werle, *Coordinating Technology: Studies in the International Standardization of Telecommunications*

Wesley Shrum, Joel Genuth, and Ivan Chompalov, *Structures of Scientific Collaboration*

Charis Thompson, *Making Parents: The Ontological Choreography of Reproductive Technology*

Dominique Vinck, editor, *Everyday Engineering: An Ethnography of Design and Innovation*

Printed in the United States
by Baker & Taylor Publisher Services